PATRIZIA GIAMPIERI-DEUTSCH (Ed.)

PSYCHOANALYSIS AS AN EMPIRICAL, INTERDISCIPLINARY SCIENCE

Collected Papers on Contemporary Psychoanalytic Research

ÖSTERREICHISCHE AKADEMIE DER WISSENSCHAFTEN
PHILOSOPHISCH-HISTORISCHE KLASSE
SITZUNGSBERICHTE, 717. BAND

VERÖFFENTLICHUNGEN DER KOMMISSION FÜR PHILOSOPHIE UND PÄDAGOGIK

Heft 27

HERAUSGEGEBEN VON HANS-DIETER KLEIN

VERLAG
DER ÖSTERREICHISCHEN AKADEMIE DER WISSENSCHAFTEN
WIEN 2005

ÖSTERREICHISCHE AKADEMIE DER WISSENSCHAFTEN
PHILOSOPHISCH-HISTORISCHE KLASSE
SITZUNGSBERICHTE, 717. BAND

PATRIZIA GIAMPIERI-DEUTSCH (Ed.)

PSYCHOANALYSIS AS AN EMPIRICAL, INTERDISCIPLINARY SCIENCE

COLLECTED PAPERS ON CONTEMPORARY PSYCHOANALYTIC RESEARCH

VERLAG
DER ÖSTERREICHISCHEN AKADEMIE DER WISSENSCHAFTEN

WIEN 2005

Vorgelegt von w. M. Hans-Dieter Klein
in der Sitzung am 15. Oktober 2004

British Library Cataloguing in Publication data.
A Catalogue record of this book is available from the British Library.

Die verwendete Papiersorte ist aus chlorfrei gebleichtem Zellstoff hergestellt, frei von säurebildenden Bestandteilen und alterungsbeständig.

ISBN 3-7001-3386-3

Satz, Layout: Robert Göschl, Universität Wien

Gesamtherstellung: Grasl Druck & Neue Medien, 2540 Bad Vöslau

http://hw.oeaw.ac.at/3386-3
http://verlag.oeaw.ac.at

Printed and bound in Austria

CONTENTS

Introduction

I. APPROACHING CONTEMPORARY PSYCHOANALYTIC RESEARCH

PATRIZIA GIAMPIERI-DEUTSCH

"[...] the intellect and the mind are objects for scientific research in exactly the same way as any non-human things. Psycho-analysis has a special right to speak for the scientific *Weltanschauung* at this point [...]"

The question of a *Weltanschauung*, 35th Lesson of *New Introductory Lectures on Psycho-Analysis* (Freud 1933 a [1932], p. 159).

In the present volume current research in psychoanalysis interfaces with research in philosophy of mind and experimental sciences, especially in cognitive neuroscience. The "common ground" of these collected essays is their deeply shared interest in an emerging science of the mind. The book contains a collection of original papers categorized under three headings: Section I is devoted to the recent research on conscious and unconscious processes in psychoanalysis, philosophy of mind and experimental sciences; Section II provides an overview of current interdisciplinary research in neuroscience and psychoanalysis and Section III explores empirical research trends in contemporary psychoanalysis. Finally, some articles in this volume can also be understood as an attempt to establish a foundation of empirical and experimental research in psychoanalysis on the quite modern epistemological approach of Freud himself.

1. PSYCHOANALYSIS, PHILOSOPHY OF MIND AND EXPERIMENTAL SCIENCES

1. 1. CURRENT ANALYTIC PHILOSOPHY OF MIND AND COGNITIVE NEUROSCIENCE

Current analytic philosophy of mind attempts to outline a theory of mind as psychoanalysis did in the past and is now doing again, but despite the seeming similarity of their aims a co-operation of the two is very challenging.

Up until now attempts to bring psychoanalysis and philosophy together have generally failed.

Philosophers who praised the particular knowledge of psychoanalysis, for example Ricœur and Habermas, managed rather to put psychoanalysis outside of the empirical sciences, whereas philosophers of science – including Ernest Nagel, Sir Karl Popper, Adolf Grünbaum[1] – did not regard it as a science at all.

Psychoanalysis is thankfully still alive, it has survived those attempts to distance it from empirical sciences quite successfully – even though in some cases unwittingly –, and it is indeed philosophy which has become increasingly threatened with being banished to so called "reservations".

A remnant of such alliances with hermeneutics or such encounters with philosophy of science has resulted in the fact that psychoanalysis does not really see itself as a science.

Current analytic philosophy of mind is no longer an "armchair philosophy" or an "armchair psychology", but it now tries to integrate the outcomes of experimental sciences as neuroscience and cognitive science, as current psychoanalysis does by updating its metapsychology.

Analytic philosophy was a philosophy of language that seriously questioned the possibility of a suitable theory of mind and even derided it as a "philosophy of ghosts". Its goal was to overcome the ambitious philosophy as metaphysics through the unpretentious analysis of language.

This modest aim became a quite pretentious one: philosophy could only analyze sentences, but at last it could analyze *every* sentence – including the sentences of experimental sciences – and decide if they are meaningful.

Analytic philosophy now raises again the old *metaphysical questions* (for example, ontological questions about main topics such as

---

[1] According to Ernest Nagel (Nagel 1959) psychoanalysis is not a science because it does not study an observable behavior, as a scientific psychology has to do, and it has no operational definitions, but it is concerned with "metaphoric" entities as so called mental phenomena are; according to Popper (Popper 1962) psychoanalysis is not a science because its sentences are not falsifyable as sentences of the other sciences indeed are; according to Grünbaum (Grünbaum 1984, 1991, 1993, 2002) – the most accurate of the three – the clinical theory of psychoanalysis is not yet a scientific one because of the lack of empirical research proving that psychoanalysis heals.

"What is consciousness? What is a representation? What is an affect?" and so on) but it requires provisional *empirical answers*, answers that only experimental sciences can provide.

Most analytic philosophers and also those I am interested in assume a naturalistic approach. The philosophical approach of naturalism is that experimental sciences are the paradigm of knowledge: to the question "What do we know about something?" naturalism tells us that our knowledge about something comes from the results of experimental sciences; and to the question "How do we get reliable knowledge?" naturalism answers that the proceedings are those of experimental sciences.

I would like to give a brief insight into this trend within the philosophy of mind.

In the current philosophy of mind the articles and books of naturalistic philosophers such as Ned Block (Block 1995), Owen Flanagan (Flanagan 1992; Güzeldere, Flanagan & Hardcastle 2000), John Searle (Searle 1992), Fred Dretske (Dretske 1995), David Chalmers (Chalmers 1996) attempt to answer, for example, what consciousness is through the interpretation of neuropsychological syndromes.

Some naturalistic philosophers have considered neuropsychological studies of brain-damaged patients with specific perceptual and cognitive deficits who have access to knowledge of which they are unaware. Brain-damaged patients own a knowledge of stimuli but they cannot perceive them with awareness.

Some neuropsychological syndromes involving dissociations between perception (mostly vision) and awareness of perception are quoted in most current philosophical works. For example, the following case histories became particularly popular: 1. blindsight patients (who have "blind" areas in their visual fields, but who however preserved visual abilities, despite no conscious awareness of perception); 2. prosopagnosics (who are unable to recognize even the faces of their own relatives, but however have "covert knowledge" of their facial expressions); 3. patients with neglect (who cannot report to the stimuli occurring on the side of space opposite to the lesion, their behavior suggests that they do not perceive stimuli, but they have non aware perception of neglected stimuli); 4. alexia patients (who can no longer read words at a glance, but who can implicitly read). All these patients have a "covert knowledge", an access to a knowledge of something without awareness so that they claim to be "guessing" (Farah 1994). Also 5. brain-split patients 6. and epileptics[2] came into vogue among philosophers of mind.

These syndromes are so interesting for philosophers because neuropsychological findings lead to a revision of the assumption that perception and consciousness of perception are always inseparably related. Philosophers are also interested in empirical issues, which help them understand the relations between the neural systems underlying perception and conscious awareness.

Therefore in the current work of philosophers of mind analysis of language and concepts are no longer so common as are case histories in order to know more about such issues as for example consciousness, its relationship with perception, and also about its physical basis.

This sketch shows that it has become evident that until now psychoanalytic case histories and, more importantly, their theoretical inferences have been left out of consideration by philosophers of mind. It shows that the common knowledge of other sciences about current psychoanalysis is too poor or too limited and that it is a necessary task, as this volume attempts to do, to bring current psychoanalytic research approaches and findings to the attention of other scientific disciplines, which pertain to the mental field (cf. also Giampieri-Deutsch 2003a, 2004a). At the same time this volume brings to attention those philosophers of mind such as David Rosenthal, Max Velmans and Pierre Jacob, who have already started to consider also psychoanalysis among the scientific disciplines of the mind.

1. 2. FREUD AND SCIENCE

By reading Freud in the light of current analytic philosophy it becomes clear how little both his critics and his followers have understood Freud's own epistemologic approach. Nowadays when psychoanalysis increasingly carries out empirical research, it is important to remember how well founded and sound Freud's account of psychoanalysis as a science was (Giampieri-Deutsch 1992, 1999, 2003b). I will substantiate my statements by reference to Freud.

---

[2] Well-known is e. g. the epileptic patient HM, studied by Brenda Milner and Larry Squire (Milner, Squire & Kandel 1998). HM had been operated and remained without his medial temporal lobe, parts of hippocampus and of amygdala in both hemispheres. He became amnestic, lost his conscious, declarative, explicit memory (memory about people, places, objects) and retained his unconscious, procedural, implicit memory (perceptual and motor skills).

1. 2. 1. Freud: was he a monist or a dualist (a psychophysical parallelist)?

I want to argue against the assumption that Freud by founding psychoanalysis abandoned the philosophical standpoint of monism in its form as physicalism and that he became a dualist.

I maintain that the position of Freud developed from a kind of physicalism, an ontological reductive physicalism, into another kind of physicalism, into a nonreductive physicalism and that he remained a monist and in so doing avoided the difficulties of dualism.

A comparison with the present debate in the current philosophy of mind allows us to understand Freud's theory of mind as being well established in this regard. I want to mention Jerry Fodor as a nonreductive physicalist who acknowledges both the mental as a legitimate object of scientific research and special sciences which are autonomous from physics (Fodor 1975).

1. 2. 2. Difficulties in holding a dualistic position

Moritz Schlick, the founder of the Vienna Circle, had already pointed out the assumption of the closed causality of the world (Schlick 1925). Current analytic philosophy of mind has shown some difficulties of dualism concerning the following assertions. They are well-sketched, in German-speaking countries, in the so-called Bieri-Trilemma (Bieri 1993, p. 9):

Mental phenomena are not physical phenomena.
Mental phenomena cause physical phenomena.
The field of the physical phenomena is closed.

The first assumption characterizes a dualistic position.

The second one is the assumption of mental causation, which ascribes causality to mental states, events, phenomena.

The third assumption states that a physical phenomenon is only explained through a physical cause.

The three assumptions taken together contradict each other and the solution of most current analytic philosophy of mind is to understand mental phenomena as a particular kind of physical phenomena. This does not mean that the body causes the mind. Mental properties or qualities emerge or supervene at a particular high level of organisation of physical phenomena they are dependent upon it, but also not reducible to it.

This is the position of nonreductive physicalism.

1. 2. 3. Freud as a monist and as a nonreductive physicalist

Freud's letter to Fließ from September 21 1897 has widely been considered as the beginning of psychoanalysis: mental events (e. g. sexual fantasies) cause mental illness (e. g. hysteria) (Freud 1985, p. 264–266). Hysteria presents not only mental, but also physical symptoms: that is mental events cause physical events. Thus – according to a widely held assumption to be refuted in the following reflections – Freud leaves physicalism and becomes a dualist.

But even in the year 1930 – 33 years after his discovery – Freud answered in a letter to Juliette Boutonier:

"I have no difficulty in acknowledging a physical world in addition to the mental world in the way that the latter [P. G.-D.: the mental world] is indeed a part of the former [P. G.-D.: the physical world]. The question of the relationship between physical and mental only applies to the latter (the mental). The physical world has a mental side in so far as it is also known by us only by means of mental perception. On the other hand our mental perceptions force us also to assume a physical reality behind mental life" (Freud 1955e [1930] translated by P. G.-D.).[3]

The psychical world was regarded by Freud as a "part", as a "side" of the physical world. This means that for Freud the mind is physically realized, embodied.

Contemporary philosophers, dualists and also monists do no longer speak of ultimate substances. If they are monists and nonreductive physicalists they refer to mental qualities in order to overcome the difficulties of the so-called Bieri-Trilemma.

---

[3] Freud wrote in his letter to Juliette Boutonier: "daß ich keine Schwierigkeit darin finde, eine physikalische Welt neben der psychischen anzuerkennen in der Art, daß die letztere [P. G.-D.: die psychische] ein Teilgebiet der ersteren [P. G.-D.: die physikalische] ist. Die Frage der Relation zwischen physikalisch und psychisch kommt nur für letztere[s] (das Psychische) in Betracht. Die physikalische Welt hat eine psychische Seit[e] insofern, als auch sie von uns nur durch psychische Wahrnehmung erkannt wird. Andererseits drängen uns unsere psychische[n] Wahrnehmungen auch die Notwendigkeit der Annahme einer physikalische[n] Realität hinter dem Seelenleben auf" (Freud 1955 e [1930], p. 672).

Freud never considered the mind as another substance, but rather as a quality, as a property. The mental is in a very modern manner a quality, as Freud wrote in "Psychical Qualities", the 4th chapter of *An Outline of Psychoanalysis* (Freud 1940a [1938], pp. 157–164).

Only now in the light of current philosophy of mind, of nonreductive physicalism, do Freud's assumptions show the novelty of his approach.

1. 2. 4. Freud was not a dualist in the form of parallelism

In *On Aphasia* Freud wrote something which was to be misunderstood as an assumption of parallelism following Jackson.

"The relationship between the chain of physiological events in the nervous system and the mental processes is probably not one of cause and effect. The former do not cease when the latter set in; they tend to continue, but, from a certain moment, a mental phenomenon corresponds to each part of the chain, or to several parts. The psychic is, therefore, a process parallel to the physiological, 'a dependent concomitant'" (Freud 1891b (1), p. 55).

In order to understand this quotation it is important to look closely at the description itself and not to be mislead by the terminology.

The sentence "The relationship between the chain of physiological events in the nervous system and the mental processes is probably not one of cause and effect" means only that there are no psychophysiological laws, as yet. Is it perhaps a question of an "anomalous Monism" as investigated by Donald Davidson? Davidson's proposal in "Mental events" (Davidson 1970) is rather charming, because he, a monist and a physicalist, finally admitted mental events and challenged the reductionism of the other analytic philosophers, but it is at the same time also rather hazardous. Indeed still many years later Davidson maintained: "The basic concepts of psychology are […] not reducible to those of physics, and therefore do not lend themselves to inclusion in a systematic, precise and closed system of description and laws. This is why, particularly when we want to talk of the relations between events described in psychological terms and events described in physical terms, we make necessary use of causal concepts; these give us the sort of forgiving conceptual connections and explanations we want" (Davidson 1993, p. 312). According to Davidson the mental is anomalous and for this reason he argued against the possibility of strict psychological laws.

However without laws there is no science, therefore according to Davidson psychology is no longer a science, but rather a philosophical discipline.[4]

When Freud noticed "from a certain moment, a mental phenomenon corresponds to each part of the chain, or to several parts" he told us that *at a particular instant* the mind supervenes to the physical or emerges from the physical.

The dependence-thesis is emphasized by the remark "a dependent concomitant". Interestingly, what appears as a quotation from Jackson cannot be found in any of Jackson's writings quoted by Freud, but is actually Freud's own words.[5] Although Jackson was a parallelist Freud did not follow him in this regard even though Freud adopted the term "process parallel".

Scholars who maintain that Freud was a parallelist seem to ignore that the dualistic theory of parallelism says that there is no causal relation at all between the mental and the physical, the two domains are entirely unrelated. Therefore mental processes and physical processes run as two clocks synchronously, completely independent of each other and subject to God as the clock-maker, as the philosopher Leibniz in his well known example stated.

Exactly the opposite is true, because the physical and the mental were regarded by Freud throughout his career from the beginning as a causally connected "continuum". In 1915, in "The unconscious", Freud could formulate this point in a very clear manner, when he referred to "the insoluble difficulties of psycho-physical parallelism" (Freud, 1915 e, 168) showing that he was fully aware of the philosophical discussion about this topic. But already in his letter to Fließ in September 22, 1898 Freud wrote:

"I am [...] not at all inclined to leave the psychology hanging in the air without an organic basis. But apart from this conviction I do not know how to go on, neither theoretically nor therapeutically, and therefore must behave as if only the psychological were under consideration" (Freud 1985, p. 326).

---

[4] For a critique of Davidson's philosophy of science see Giampieri-Deutsch (2004b, pp. 84–87).

[5] Cf. footnote 2 of Paul Vogel, the editor of the original German edition *Zur Auffassung der Aphasien*. Frankfurt am Main: Fischer 1992 (Freud 1891b (2), p. 98).

Throughout the "Project for a scientific psychology" monism is emphasized as a unity, an integrated whole of the physical and the mental (Freud 1950 c [1895]).

In *Three Essays on Sexuality* drives are representatives of the somatic element which is the basis of the mind: "By an 'instinct' is provisionally to be understood the psychical representative of an endosomatic, continuously flowing source of stimulation [...] The concept of instinct is thus one of those lying on the frontier between the mental and the physical. The simplest and likeliest assumption as to the nature of instincts would seem to be that in itself an instinct is without quality, and, so far as mental life is concerned, is only to be regarded as a measure of the demand made upon the mind for work" (Freud 1905 d, p. 168).

Freud did not abandon the view that psychoanalysis would find a link to neurobiology: "The theoretical structure of psycho-analysis that we have created is in truth a superstructure, which will one day have to be set upon its organic foundation" (Freud 1916–17 a [1915–17], p. 388).

As he later insisted in *The Question of Lay Analysis*: "In view of the intimate connection between the things that we distinguish as physical and mental, we may look forward to a day when paths of knowledge and, let us hope, of influence will be opened up, leading from organic biology and chemistry to the field of neurotic phenomena" (Freud 1926 e, p. 231).

1. 2. 5. Freud on epiphenomenalism

In the section "The consciousness" in the first part of the "Project for a scientific psychology" Freud commented upon other theories of consciousness of his time: "According to an advanced mechanistic theory, consciousness is a mere appendage to physiologico-psychical processes and its omission would make no alteration in the psychical passage [of events]." Indeed Freud remarked upon the epiphenomenalism of Huxley. Freud argued against epiphenomenalism because this approach rejected the possibility of mental causation and instead explained his own theory: "Here consciousness is the subjective side of one part of the physical processes in the nervous system, namely of the Ω [omega] processes; and the omission of consciousness does not leave psychical events unaltered but involves the omission of the contribution from Ω [omega]" (Freud 1950c [1895], p. 311).

1. 2. 6. Freud and causality

Scientific thought is characterized by the assumptions of both dependence and determination, in a word causality. The world is not an accumulation of unrelated phenomena, events, objects, but it constitutes a system and this conception of the world, the scientific *Weltanschauung* shared by Freud but also by the Vienna Circle (Giampieri-Deutsch 1997), the interrelatedness of things is the basis of our scheme of things. A mental or a physical thing is dependent on another thing or is determined by other things. This then brings about scientific knowledge i. e. explanation, prediction and control of events. Determination is not determinism. Determination is a concept which belongs to the theory of science, determinism is a concept which belongs to the ideology of science.

By taking this seriously Freud had to assume the existence of the unconscious: "the equation of what is mental with what is conscious had the unwelcome result of divorcing psychical processes from the general context of events in the universe and of setting them in complete contrast to all others" (Freud 1940 b [1938], p. 283).

In *The Interpretation of Dreams* Freud pointed out his contemporaries' distrust in and at the same time his own trust in causality: Freud was dealing here with the question of mental causation: "It is true that the dominance of the brain over the organism is asserted with apparent confidence. Nevertheless, anything that might indicate that mental life is in any way independent of demonstrable organic changes or that its manifestations are in any way spontaneous alarms the modern psychiatrist, as though a recognition of such things would inevitably bring back the day of the Philosophy of Nature, and the metaphysical view of the nature of mind. The suspicions of the psychiatrists have put the mind, as it were, under tutelage, and they now insist that none of its impulses shall be allowed to suggest that it has any means of its own. *This behaviour of theirs only shows how little trust they really have in the validity of a causal connection between the somatic and the mental. Even when investigation shows that the primary exciting cause of a phenomenon is psychical, deeper research will one day trace the path further and discover an organic basis for the mental event.* But if at the moment we cannot see beyond the mental, that is no reason for denying its existence" [Emphasis of P. G.-D.] (Freud 1900a, pp. 41–42).

1. 3. PSYCHOANALYSIS AS A SCIENCE

From the beginning to the later writings and even to the very last posthumously published manuscripts Freud looked at psychoanalysis constantly, continuously, and persistently as a natural science. Was it a "scientistic self-misunderstanding" as Habermas argued in his well-known book *Knowledge and Human Interest* (Habermas 1968, p. 246)? In this regard Grünbaum defended Freud's epistemologic position against Habermas' interpretation (Grünbaum 1984, p. 8).

Freud explained in his posthumously published manuscript "Project for a scientific psychology" which he intended to become a "psychology for neurologists" as he wrote in the letter to Fließ of April 27 1895): "The intention is to furnish a psychology that shall be a natural science: that is, to represent psychical processes as quantitatively determinate states of specifiable material particles [...]" (Freud 1950c [1895], p. 295).

Not long after this Freud gave up trying to reduce clinical theory to neurobiology. He abandoned ontological reduction in favour of a nonreductive methodologic physicalism.

The ontological reduction of the ontological physicalism, the first old fashioned version of physicalism, which was the account of Helmholtz, Du Bois-Reymond and of Freud's teacher Brücke, stated that everything which can be studied scientifically is a physical thing. Freud rejected this approach as he pointed out for example in "Psychical (or mental) treatment": "After a somewhat unfruitful period during which it [P. G.-D.: medicine] was dependent on what was known as "Natural Philosophy", it came under the happy influence of the natural sciences and has achieved the greatest advances alike as a science and as an art [...] All of these advances and discoveries were related to the *physical* side of man, and it followed, as result of an incorrect though easily understandable trend of thought, that physicians came to restrict their interest to the physical side of things and were glad to leave the mental field to be dealt with by the philosophers whom they despised" (Freud 1890a, pp. 283–284).

But Freud also rejected the theoretical reduction of the theoretical physicalism or mechanism which is a second account of physicalism stating that the laws of every science are derived from or can be reduced by the theories of physics. An example in "Resistances to psychoanalysis" illustrates how clearly Freud differentiated between physicalism and mechanism: "During the materialistic or, rather, mechanistic period, medicine made tremendous advances, but it also showed a short-sighted misunderstanding of the most important and most difficult among the problems of life" (Freud 1925e [1924], p. 216).

Physicalism does not have to be a form of theoretical reductionism and does not have to take physics as a model as mechanism does. This second version of physicalism does not deny the reality of mental phenomena, but devalues the theories of other sciences, so named "special sciences" by Ned Block and Jerry Fodor, such as psychology, sociology, economics, etc., the theories of which can not be derivable from the theory of physics (cf. Giampieri-Deutsch 2004b). The claim is that concepts of scientific theories have to be reduced by means of "operational definitions" to proceedings whose outcomes can be tested through observation.

Freud committed himself to a third version of physicalism, a very modern, for his time revolutionary and until now a little understood approach, the methodologic physicalism. Freud's nonreductive methodologic physicalism claimed rather that science discovers regularities in the relationship between phenomena, studies their causal relations and proposes generalizations, laws and explanations. All scientific knowledge is provisional and is subject to testing and control (Freud 1933a [1932], pp. 172–173). According to Freud every science has to meet this standard.

1. 3. 1. But why did Freud speak of 'natural science' and not just of 'science'?

Freud's impressive knowledge of philosophy of science was derived from his teacher Franz Brentano, as I maintained in several of my studies on Freud's relationship with philosophy. Brentano is generally known for his theory of intentionality. As a Darwinist, as a methodologist, and as the grandfather of the Vienna Circle he is not so well-known. For Brentano natural science was synonymous with empirical science, but this does not entail a reduction to physics. Brentano wrote in 1866 as the 4th thesis for his *venia docendi*: "The true method of philosophy is none other than that of natural science."[6] At the beginning of

[6] "Vera philosophiae methodus nulla alia nisi scientiae naturalis est" (Brentano 1924, p.136).

his *Psychology from an Empirical Standpoint* he stated: "My place in psychology is at the empirical viewpoint. My only master is experience" (Brentano 1874, p. 1).

As Freud himself stressed in his *An Autobiographical Study* he always regarded it as an unfairness to psychoanalysis that it had not been considered in the same way as any other natural science: "I have always felt it as a gross injustice that people have refused to treat psychoanalysis like any other science" (Freud 1925d [1924], p. 58).

As Freud later explained in "Some elementary lessons in psychoanalysis" about "The nature of the psychical": "Psycho-analysis is a part of the mental science [*Seelenkunde*] of psychology. [...] Psychology, too, is a natural science. What else can it be?" (Freud 1940b [1938], p. 282).

In the 35th lesson of *New Introductory Lectures on Psycho-Analysis* "The question of a Weltanschauung" in the year 1933 Freud leaves no doubt about his own epistemologic position on this subject: "the intellect and the mind are objects for scientific research in exactly the same way as any non-human things. Psycho-analysis has a special right to speak for the scientific *Weltanschauung* at this point, since it cannot be reproached with having neglected what is mental in the picture of the universe. Its contribution to science lies precisely in having extended research to the mental field. And, incidentally, without such a psychology science would be very incomplete. If, however, the investigation of the intellectual and emotional functions of men (and of animals) is included in science, then it will be seen that nothing is altered in the attitude of science as a whole, that no new sources of knowledge or methods of research have come into being" (Freud 1933a [1932], p. 159).

In the posthumously published *An Outline of Psycho-Analysis* Freud concluded: "Whereas the psychology of consciousness never went beyond the broken sequences which were obviously dependent on something else; the other view, which held that the psychical is unconscious in itself, enabled psychology to take its place as a natural science like any other. The processes with which it is concerned are in themselves just as unknowable as those dealt with by other sciences, by chemistry or physics, for example; but it is possible to establish the laws which they obey and to follow their mutual relations and interdependences unbroken over long stretches – in short, to arrive at what is described as an 'understanding' of the field of natural phenomena in question" (Freud 1940a [1938], p. 158).

1. 3. 2. Concepts in psychoanalysis, as in every other science, are indeterminate and not invariable

In *An Outline of Psycho-Analysis*, already referred to above, Freud pointed out that "the basic concepts and principles of the new science (instinct, nervous energy, etc.) remain for a considerable time no less indeterminate than those of the older sciences (force, mass, attraction)" (Freud 1940a [1938], p. 159).

Metapsychological basic concepts like those of the theory of mind are not invariable, unchangeable as a long tradition beginning with Robert Waelder (Waelder 1962) and Benjamin B. Rubinstein (Rubinstein 1967) has believed it to be. These concepts – as Freud stated very clearly in this regard from the point of view of current philosophy of science – "can lay claim to the same value as approximations that belongs to the corresponding intellectual scaffolding found in other natural sciences, and we look forward to their being modified, corrected and more precisely determined as further experience is accumulated and sifted" (Freud 1940a [1938], p.159).

By the way, Freud unmistakably expressed his own view on this topic even as late as 1925 in his autobiography: "Clear basic concepts and sharply drawn definitions are only possible in the mental sciences [*Geisteswissenschaften*] in so far as the latter seek to fit a region of facts into the frame of a logical system. In the natural sciences, of which psychology is one, such clear-cut general concepts are superfluous and indeed impossible. [...] The basic ideas or most general concepts in any of the disciplines of science are always left indeterminate at first and are only explained to begin with by reference to the realm of phenomena from which they were derived; it is only by means of a progressive analysis of the material of observation that they can be made clear and can find a significant and consistent meaning" (Freud 1925d [1924], pp. 57–58).

2. "FIRST-PERSON PERSPECTIVE" AND "THIRD-PERSON PERSPECTIVE" IN PSYCHOANALYSIS

I will apply both terms "first-person perspective" and "third-person perspective" to psychoanalysis as they are understood and used in current analytic philosophy of mind. I will not be considering some influential classical linguistic investigations of analytic philosophy. For instance Donald Davidson confined the problem of "first-person perspective" into a narrower frame as "self-attribution" and "self-ascription", by considering it, the "first-person perspective", "only as it applies to propositional attitudes like belief, desire, intention" (Davidson 1984, p. 3). Although such studies are – now as before – seminal, I would like to emphasize some new aspects of the concepts "first-person perspective" and "third-person perspective", which have arisen from the current debate in order to point out: first, the kind of knowledge and data provided by clinical research and empirical research in psychoanalysis; and second, that the particular knowledge of psychoanalysis in interdisciplinary dialogue with related fields can provide answers to some of the questions of other disciplines, for instance to the "other minds problem" of philosophy of mind.

Within the classical analytic philosophy the "first-person perspective" was originally the answer to the question about the access to the knowledge of one's own mind. Direct knowledge, privileged access, privacy, first-person authority, incorrigibility and self-evidence were some of the classical epistemological criteria, which analytic philosophers elaborated on.

The perspectivity of these points of view includes the epistemic asymmetry in the investigation of mind. The view from the "third-person-perspective" is a view "from outside", but the view from the "first-person perspective" is not only "from inside", but also "for oneself". Current analytic philosophy of mind uses the term first-person perspective connected to or even interchangeably with terms such as subjectivity, subjective experience, "what it is like to be" mental state, "raw feels", phenomenal consciousness, subjective awareness, phenomenal properties, qualia, qualitative character of experience and a variety of others so that the "first-person perspective" can at best be defined *ostensively* (Güzeldere 1997, pp. 42–43; Searle 1992, pp. 83ff.). In front of the linguistic analysis of classical analytic philosophy the current *ostensive definition* of the term "first-person perspective" in philosophy of mind may be criticized as circular, it seems that, as Ned Block notices, "really all one can do is *point* to the phenomenon" (Block 1995, p. 380). Some philosophers matintain that to refer the phenomenal aspect of the mind exclusively to sensory properties is a misunderstanding, as Robert van Gulick states: "Phenomenal experience is not merely a succession of qualitatively distinguished sensory ideas, but rather the organized cognitive experience of a world of objects and of ourselves as subjects within that world" (Van Gulick 1993, p. 137).

2. 1. Some objections from Continental and analytic philosophy as well as from the common sense of the psychoanalysts

Psychoanalysis is the science, which integrates the "first-person perspective" in its own approach to the investigation of mental states (Giampieri-Deutsch 2003b, p. 64). However was it not exactly Freud who unveiled the deceptive inner perception of his patients and pushed forward to their unconscious thoughts and threatened the status of one's knowledge?

So at first glance my definition of psychoanalysis as the science which integrates the "first-person perspective" may sound quite challenging. There are at least two philosophical traditions which both seem to question this statement, Continental philosophy and analytic philosophy. Furthermore we should consider the everyday clinical experience of my colleagues and of myself, who have to question the first-person authority of the patients.

Concerning Continental philosophy Paul Ricœur defined Freud – together with Nietzsche and Marx – as one of the "masters of suspicion" (Ricœur 1965), who taught us that the first-person authority misleads us and should be interpreted. Michael Foucault also argued in a similar vein in 1967 that after Freud – again connected with Nietzsche and Marx – "l'interprétation est enfin devenue une tâche infinie" (Foucault 1967, p. 187).

Among the analytic philosophers Joseph Agassi showed a similar understanding of Freud: "The doctrine of privileged access is that I am the authority on all my own experiences [...] The thesis was refuted by Freud (I know your dreams better than you) [...]" (Agassi 1975, p. 120) Everyday clinical experience shows us psychoanalysts the illusions of the authority of the "first person perspective", for instance in the transference. A tentative definition of the mark of psychoanalysis among the other psychotherapies is, that its major concern is about detecting and interpreting transference. Additionally transference is exactly the mental (and bodily) phenomenon in which the authority of the "first-person perspective" seems dramatically to fail.

These objections weaken, if the concept of the "first-person perspective" is understood more widely.

A report given in first-person terms generally describes an experience in terms of how it feels to the subject who experiences it (cf. Farrell 1950; Nagel 1974; Jackson 1982, 1986; Shoemaker 1993). In the current debate the "first-person perspective" is no longer seen as a sort of Cartesian transparency (a special inner perception). The discussion in analytic philosophy of mind addresses a very crucial topic. Although psychoanalysis questions the "first-person perspective", this perspective is absolutely necessary for it. Psychoanalysis has always known what current analytic philosophy of mind is now discovering: mental states of the first-person have not only a cognitive content, they also feel to the subject in a particular way[7]. Mental states of the first-person allow a particular epistemic access to some "what it is like to be" states not least through intersubjective reflective working through. Otherwise the psychoanalytic clinical method, which is based on free associations and insight of the analysand, would not be possible.

In current analytic philosophy of mind the "first-person perspective" is often misunderstood as a solipsistic and unknowable mental state, but psychoanalytic experience provides some advances in the philosophical understanding of this matter.

Are the "first-person perspective" and the "third-person perspective" mutually exclusive? Max Velmans' contribution in this volume addresses this point (see below). In my opinion too, psychoanalysis seems to be the best living proof of an integration of both perspectives in a science.

2. 2. "First-person perspective" and "third-person perspective" in clinical psychoanalytic research

Psychoanalytic clinical method provides a systematic exploration of subjective experience and is a source of clinical data about the mind. In the clinical situation the analyst knows from the "first-person perspective" in evenly hovering attention and in the phenomena of countertransference and projective identification. However the analyst knows also from the "third-person perspective" on two different levels: on the first level in clinical work and research by writing case histories, by participating in supervision, in intervision, or in case history seminars. There is also a second level of the "third-person perspective" the psychoanalytic empirical research, which claims more compelling "objectivity" and which I will consider later.

---

[7] Wittgenstein wrote in § 302 of his Philosophical Investigations "If one has to imagine someone else´s pain on the model of one´s own, this is none too easy a thing to do: for I have to imagine pain which I do not feel on the model of the pain which I do feel" (Wittgenstein 1953, p. 101).

I would like to introduce the concept of projective identification. It is the way in which a subject, the patient, makes use of another subject, the analyst, to experience aspects of himself. In this case the analyst is pressured to experience and feel – including bodily – in a way that is congruent with the experience of the patient (cf. Ogden 1982, pp. 22–23). Sándor Ferenczi discovered it as early as 1911 as "'*L' autre danger*' that the psychoanalyst is subject to [...] Besides monitoring the countertransference, one must therefore also pay heed to this 'being induced' by the patients. (Perhaps it is only a question here of a form of countertransference)." as Ferenczi (letter 198 Fer) wrote to Freud on February 1911 (Freud & Ferenczi 1993, p. 253; cf. Giampieri-Deutsch 1996). Today one cannot say that the concept projective identification "belongs" to any particular school, but Melanie Klein officially introduced it in "Notes on some schizoid mechanism" (Klein 1946). A very important point made by Melanie Klein is that a subject projects not "onto" but rather "into" another person[8]. The analyst is not only a mirror in which the internal objects of the patients are reflected in the transference. The patient projects, "puts" his or her feelings and fantasies not only *onto* the analyst but also *into* him or her.

According to experienced clinicians such as Bott-Spillius: "patients can behave in ways that get analysts to feel the feeling that the patient, for one reason or another, cannot contain within himself or cannot express in any other way except by getting the analyst to have the experience too" (Bott-Spillius 1992, p. 61). "The analyst's aim is to allow himself or herself to experience and respond internally to such pressures and of its contents so that he can interpret it, but without being pushed into gross acting out. Some degree of acting out by the analyst, however, is often inevitable in the early stage of what the patient is feeling [...]" (Bott-Spillius 1992, p. 63). The clinical experience also teaches that "the analyst is always affected to some degree by his patient's projections" (Bott-Spillius 1992, p. 64).

A phenomenon experienced from the "first-person perspective" in everyday clinical life becomes in psychoanalytic clinical research a phenomenon observed from the "third person perspective" and therefore a result. According to Ogden: "Unlike many of the 'beliefs' of the different schools of psychoanalytic thought, projective identification is not

[8] "Together with these harmful excrements, expelled in hatred, split-off parts of the ego are also projected onto the mother, or, as I would rather call it, into the mother" (Klein 1946, p. 8).

a construct that one accepts or rejects on the basis of an attraction to a metaphor [...] Projective identification is a clinical-level conceptualization with three phenomenological references, all of which lie entirely within the realm of observable psychological and interpersonal experience: (1) the projector's unconscious fantasies (observable through their derivatives, such as associations, dreams, parapraxes, and so forth); (2) forms of interpersonal pressure that are often subtle but verifiable; and (3) countertransference experience (a real, yet under-utilized source of analysable data)" (Ogden 1982, pp. 45–46).

Psychoanalytic clinical findings as well as their theoretical inferences have not yet been considered enough by philosophers of mind and scientists. The common knowledge of related sciences about current psychoanalysis is still quite limited and current psychoanalytic results should be shared with related sciences and disciplines. But "clinical data" is held to be "weak data". For this reason psychoanalytic empirical research tries to provide "empirical data", which can be considered "hard data" form the points of view of both experimental sciences and philosophy of mind.[9]

Until now current discussion in analytic philosophy of mind related "first-person perspective" to consciousness. Nonetheless from the comparison between philosophy of mind and clinical psychoanalytic experiences some questions arise: For example, should the "first-person perspective" be considered as a conscious state from a psychoanalytic point of view? Or can it also be conscious, preconscious or unconscious? David Rosenthal, a philosopher of mind who is also committed to psychoanalysis, also addresses this question in his contribution to this volume and argues that phenomenal states can be conscious, but they don't necessarily have to be conscious (see below). Lastly, should the "first-person perspective" be considered merely a mental state or rather a psychophysical state?

2. 3. The particular knowledge of psychoanalysis, which can provide answers to some questions in other disciplines

Psychoanalysis can, in interdisciplinary dialogue and mutual exchange, learn a lot from related fields as cognitive psychology, cognitive neuroscience, neurobiology, philosophy of mind, linguistics and meth-

[9] Concerning the experimental testing of Freud's theory and clinical method see Fisher & Greenberg (1985, 1996, 2002).

odology. This present volume shows several interdisciplinary contributions to the growth of psychoanalysis as a theory of mind, as a clinical theory, and as a discipline, which is active in both clinical and empirical research.

I would like to point out, that there are a lot of obvious clinical phenomena and observations, which every psychoanalyst is aware of and that could contribute to solving what is supposed to be a "mystery" in other related disciplines. The rich common knowledge of the psychoanalyst practitioner is often undervalued by himself or by herself.

2. 4. The "other minds problem": how can we know the minds of others?

The old version of the problem emphasized the cognitive aspect. In § 243 of his *Philosophical Investigations* Wittgenstein elaborated on the "first-person perspective" and its epistemological criterion of privacy: "But could we also imagine a language in which a person could write down or give vocal expression to his inner experiences – his feelings, moods, and the rest – for his private use? – Well, can't we do so in our ordinary language? – But that is not what I mean. The individual words of this language are to refer to what can only be known to the person speaking; to his immediate private sensations. So another person cannot understand the language" (Wittgenstein 1953, p. 89).

The connection between the authority of the "first-person perspective" and the traditional problem of "other minds" was clearly formulated by Wittgenstein: "I cannot know what is going on in him" (Wittgenstein 1953, p 223). And: "What is internal is *hidden* from us" (Wittgenstein 1953, p. 223).

Wittgenstein was also concerned with the possibility of knowing other minds by "guessing" or by "reading": "'What anyone says to himself within himself is hidden from me' might of course also mean that I can for the most part not *guess* it, nor can I read it off from, for example, the movements of his throat (which would be a possibility)" (Wittgenstein 1953, p 221). In the next pages Wittgenstein insisted: "There is a game of 'guessing thoughts'" (Wittgenstein 1953, p 223).

A classical tentative solution was to argue, that behavior allows us to know other minds, because behavior was held to be observable. In his *Philosophical Investigations* Wittgenstein wrote for instance: "A doctor asks: 'How is he feeling?' the nurse says: 'He is groaning.' A report on his behaviour." (Wittgenstein 1953, p 179) As Davidson remarks: "Since Wittgenstein it has become routine to try to relieve worries about 'our knowledge of other minds' by remarking that it is an essential aspect of our use of certain mental predicates that we apply them to others on the basis of behavioral evidence but to ourselves without benefit of such aid" (Davidson 1987, p. 16).

The new version of the problem emphasized the phenomenal aspect (sensations, emotions, bodily feelings). In Thomas Nagel's text "What is it like to be a bat?" from 1974 the perspectivity of the "first-person" is directly understood as a subjective feeling of the mind, as a first-person feeling, as a "what it's like to be" feeling: "the fact that an organism has conscious experience *at all* means, basically, that there is something it is like to *be* that organism" (Nagel 1974, p. 519). It is the phenomenal facet which is focused on here. The subjective character of the experience can be grasped from only one point of view, from the first-person perspective: "we believe that these experiences also have in each case a specific subjective character, which it is beyond our ability to conceive" (Nagel 1974, p. 521).

Among the current philosophers of mind the naturalist Owen Flanagan, who is quite sympathetic to psychoanalysis, solves the "other minds problem" with an intuitive argument: "The metaphor of us as mental detectors is a powerful one. Just as a metal detector detects metal objects beneath the sand or between blades of grass, so too mental detectors detect invisible mental states. Good mental detectors understand a great deal about the form of life in which they live and about behavioral regularities" (Flanagan 1992, pp. 102–103). "Persons understand each other, but not by penetrating each other's being, not by having each other's experiences. Understanding another involves conceiving of the other's experiences" (Flanagan 1992, p. 106). Flanagan concludes: "There are other minds, and we have knowledge of them" (Flanagan 1992, p. 107). According to Flanagan there is no reason why a naturalist cannot both acknowledge the existence of subjectivity and the subjective "first-person perspective" by getting a description of them as a part of the general project of understanding the mind. The philosopher of mind Pierre Jacob discusses findings of cognitive science on mindreading in his contribution to this volume (see below).

I would also like to present some possible solutions to the "other-minds problem" offered by psychoanalytic clinical research and by psychoanalytic empirical research.

2. 4. 1. A tentative solution by psychoanalytic clinical research: projective identification

One can know other minds by everyday projective identification. This mechanism, obtained from clinical psychoanalytic research, is not restricted to psychopathology. Projective identification can be generalized to everyday life situations. León Grinberg described in "Countertransference and projective counteridentification", how projective identification – a phenomenon discovered in a clinical situation – occurs as "normal" phenomenon in everyday life: "Under normal conditions projective identification determines the empathic relationship with the object, not only because it allows one to put oneself in the place of the other and therefore understand his feelings better, but also for what it brings out in that person. The subject produces some emotional reaction in the object. […] This usually happens, within certain limits, in all human relationships and forms the basis of the communication" (Grinberg 1979, p. 170).

2. 4. 2. A tentative solution by psychoanalytic empirical research: mentalization

A further solution to the "other minds problem" is provided by psychoanalytic empirical research. Peter Fonagy's empirically rooted theory of mentalization refers to the studies of Baron-Cohen and colleagues (Baron-Cohen 1995; Baron-Cohen, Tager-Flusberg & Cohen 1993) and offers an excellent answer to mindreading: "Reflective function, or mentalization, enables children to 'read' other people's minds. By doing this, children make people's behavior *meaningful* and predictable" (Fonagy, Gergely, Jurist & Target 2002, p. 24). Mentalizing, a central aspect of social cognition, entails implicitly or explicitly perceiving or interpreting the behavior of others and oneself as intentional, i.e., mediated by mental states and processes. This developmental acquisition permits not only response to another person's behavior but to his "beliefs, feelings, attitudes, desires, hopes, knowledge, imagination, pretense, deceit, intentions, plans, and so on" (Fonagy, Gergely, Jurist & Target 2002, p. 24). According to Fonagy, making sense of minds is central to the diagnostic work of psychoanalysis and the experience of another human being. Having one's mind in mind is central to the impact of therapies, ranging from cognitive therapy to psychoanalysis. Fonagy relates mentalization to clinical concepts such as "introspection", "observing ego", "insight", and "empathy".

I would like to summarize some central thoughts of Fonagy about mentalization, which are relevant to the topic under discussion. Reflective functioning refers to the mental processes underlying the capacity to mentalize. It expresses the mental capacity related to the representation of the self and involves a self-reflective and interpersonal component. Reflective functioning or mentalizing is the "capacity to perceive and understand oneself and others in terms of mental states (feelings, beliefs, intentions and desires) [...] to reason about one's own and others' behavior in terms of mental states, i.e. *reflection*" (Fonagy, Target, Steele & Steele, 1998, p. 7). According to Fonagy, reflective functioning works unconsciously like an automatic procedural knowledge of the mind. Therefore Fonagy distinguishes reflective functioning from introspection or self-reflection, which are restricted to consciousness and are rather declarative self-knowledge or self-report, whereas reflective functioning refers to the ability to give sense to a behavior and to adjust it. The reflective function can be measured by the Reflective Functioning (RF) Scale, designed by Fonagy and colleagues (Fonagy, Target, Steele & Steele 1998) for and currently applied to the protocols of the Berkeley Adult Attachment Interview (AAI) (George, Kaplan & Main 1996).[10] The Reflective Functioning (RF) Scale is mainly implemented in different designs of psychoanalytic outcome studies to investigate the effects of psychoanalytic treatment on the improvement of the capacity to understand one's own and others' mental states. But Fonagy and colleagues provide more than an impressive theory, by substantiating it in two ways: First, if "reading" other minds is no longer a mystery but a common capacity acquired in normal development, the development of this ability can be observed and investigated by both psychoanalytic developmental psychology and cognitive psychology; second, this capacity can not only be observed, but also be assessed by the Reflecting Functioning Scale. What makes the phenomenon of knowing other minds understandable is the attention in psychoanalytic clinical and empirical research to both cognitive and emotional aspects.

A paraphrase of a famous sentence of Wittgenstein (Wittgenstein 1953, p. 222) provides a commentary: "A whole cloud of philosophy condensed into a drop of psychoanalysis."

---

[10] The Adult Attachment Interview (AAI) is a semistructured interview asking for the experiences of separation, illness, rejection and loss experienced by the interviewees in relation to the childhood caregivers, in order to assess the quality and the degree of their attachment to the significant others.

3. PSYCHOANALYSIS, PHILOSOPHY OF MIND AND EXPERIMENTAL SCIENCES: RESEARCH IN CONSCIOUS AND UNCONSCIOUS PROCESSES

In "Toward a theory on consciousness based on recent developments in subliminal research" Howard Shevrin holds that one of the most important questions in contemporary psychology and neuroscience concerns the relationship between conscious and unconscious processes. According to Shevrin consciousness cannot be fully understood without a concomitant understanding of unconscious processes, and unconscious processes cannot be fully explained without an understanding of conscious processes. Shevrin maintains that this conjoint understanding can only be achieved by placing individuality at the heart of the solution. Consciousness theorists claim that subjectivity is the hallmark of consciousness. Shevrin points out that subjectivity is always individual. Shevrin argues that there cannot be a completely free standing theory of consciousness without taking the unconscious into account, or a theory of the interaction between conscious and unconscious processes without taking individual differences into account. According to Shevrin any quantitative or qualitative differences between conscious and unconscious processes can best be seen through the lenses of individual differences. Therefore research methods should study conscious and unconscious processes jointly as a function of these individual differences. Shevrin describes a number of studies that illustrate this position in order to support empirically the following four propositions by experimental findings: 1. consciousness cannot be fully understood without knowledge of unconscious processes; 2. unconscious processes are complex, representational, highly interactive with conscious processes and not dispositional or latent; 3. the interaction between conscious and unconscious processes can best be understood when person and context are taken into account; 4. there are quantitative and qualitative differences between conscious and unconscious processes. These propositions taken together make possible a different approach to conceptualizing the nature of consciousness and its relationship to unconscious processes from that currently prevailing in much research and theorizing in he field.

In "Drive theory and primary process: a philosophical account", the psychoanalyst and psychoanalytic researcher Linda Brakel describes and defines drives. According to Freud, drives are a boundary concept, a mental representative of a state of physiological need. Each drive has a source, an aim, and an object through which satisfaction can be obtained. Brakel carries out a philosophical analysis of drives. Drives are shown by Brakel not to be adequately explained in the analyses of standard propositional-attitude psychology. This is due to the fact that drives not have singular determinate objects. The non-singularity and high displace-ability of drive aims is taken up in contrast to the indeterminacy threatening representations of drive objects. Compared to aims, drive objects have no original singular object; instead the object of drives seem to be a primary process organized set of objects. Since standard means will not work for individuating representations of this nature, Brakel raises the question concerning what can save these non-singular, primary process set type drive objects from being unconstrained and indeterminate. According to Brakel, a proper function naturalistic account of drive objects will make this possible, and thus a case is made for such primary process sets of objects having a proper function. For the proper function argument to be effective, it must be shown that such objects have a proper function under conditions which will enhance selective fitness, and this too is highlighted by Brakel. Finally, the objects of drives and their representations are contrasted, first with another sort of representation that can be individuated in a proper-function account, and then with representations understood philosophically as "vague" concepts.

The philosopher of mind and cognitive psychologist Max Velmans' contribution "Complementary first- and third-person approaches to the investigation of mind" presents his own view that, although first-person and third-person accounts of mind are not reducible to one another, they are not exclusive, but rather complementary. Psychosomatic medicine assumes that the conscious mind can affect body states, and this is supported by evidence that the use of imagery, hypnosis, biofeedback and other 'mental interventions' can be therapeutic in a variety of medical conditions. However, there is no accepted theory of mind/body interaction and this has had a detrimental effect on the acceptance of mental causation in science, philosophy and in many areas of clinical practice. Velmans maintains that the nature of mind can be partly revealed by mind/body interactions and particularly by the causal interactions of consciousness and brain. However, causal interactions between mind and body are not easy to understand. According to dualist-interactionists, brains and conscious experiences are entirely different kinds of substance that causally interact, but how they do so is not clear. According to physicalist reductionists conscious experiences are merely brain states, which explains their interaction with other brain states. However Velmans holds that reductionists habitually confound causation, correlation and ontological identity. Finding the neural causes and correlates of conscious experiences would not reduce those experiences to states of the brain. Biomedical accounts typically translate the effects of mind into the effects of brain functioning, for example, explaining mind/body interactions in terms of the interconnections and reciprocal control of cortical, neuroendocrine, autonomic and immune systems. While such accounts are instructive, they are implicitly reductionist, and beg the question of how *conscious experiences* could have bodily effects. If conscious states cannot be reduced to brain states, on the other hand the function of consciousness also poses problems, and according to Velmans non-reductionist accounts have to cope with three problems: First, the physical world appears causally closed, which would seem to leave no room for conscious intervention. Second, one is not conscious of one's own brain/body processing, so how could there be conscious control of such processing? Third, conscious experiences appear to come too late to causally affect the mental processes to which they most obviously relate. There are in any case different senses in which a mental process can be said to "be conscious". It might be conscious (a) in the way that we can become conscious *of* it, (b) of that it *results* in a conscious experience, and (c) in the sense that consciousness *causally affects* that process. Some processes are conscious in the sense of (a) or (b). However, it is uncertain whether any process is conscious in the sense of (c). Velmans points out that Freud arrived at a similar analysis of how the operations of unconscious mind relate to the contents of consciousness and had similar doubts about the causal efficacy of consciousness. However, this produces a paradox: viewed from a third-person perspective, consciousness seems unnecessary to the operations of mind, but viewed from a first-person perspective, it seems to be central to mental life. Velmans suggests a way to think about consciousness, unconscious mind and brain that allows this paradox to be resolved.

According to the philosopher of mind David Rosenthal in "Consciousness, interpretation, and higher-order-thought", a mental state is conscious only when the individual who is in this state is in some way conscious of being in that state. Rosenthal poses a central question for a theory of consciousness, which consists of what way of being conscious of a mental state constitutes its being actually conscious. Rosenthal argues that we are conscious of those states by having thoughts to the effect that we are in them. These *higher-order thoughts* are themselves seldom conscious, and they must occur independently of conscious inference. This model applies not just to cognitive and desiderative states, but equally to qualitative states, such as bodily and perceptual sensations and affective states. The model is supported by our being conscious of a given mental state in respect to different mental properties, and by confabulatory mental states, which we subjectively seem to be in, despite evidence that we are not. Rosenthal's model also helps to explain various connections between consciousness and speech. Finally, because one's having a suitable higher-order thought amounts to interpreting oneself as being in that state, a mental state's being conscious is a matter of one's noninferential self-interpretation.

The philosopher of mind and cognitive scientist Pierre Jacob presents in his contribution "First-person and third-person mindreading" some further current answers to the "other-minds problem" provided by philosophy of mind and cognitive science and offers a framework to better understand what Fonagy means by reflective functioning or mentalizing. According to Jacob, who like Fonagy refers to the developmental cognitive research of Baron-Cohen (Baron-Cohen 1995), mindreading is the cognitive ability to recognize the presence, and retrieve the contents, of human minds. First-person mindreading is knowledge of one's own mind. Third-person mindreading is knowledge of other minds. Jacob questions a recent influential proposal, which maintains that all forms of mindreading are accomplished by means of a simple cognitive heuristic: mental simulation. Jacob first argues that full-blown third-person mindreading lies outside the scope of mental simulation. Lastly, he maintains that mental simulation cannot even begin to provide an account of first-person mindreading.

4. CURRENT INTERDISCIPLINARY RESEARCH: "FIRST-PERSON PERSPECTIVE" AND "THIRD-PERSON PERSPECTIVE" IN NEUROSCIENCE AND PSYCHOANALYSIS

Psychoanalysis is the science which integrates the "first-person perspective" in its own approach to the investigation of mental states and at the same time attempts to understand subjective experience from the perspective of science. The "third-person perspective" is what is operant in the everyday scientific practice of experimental sciences and provides valuable advances in the understanding of the mind. The psychoanalyst wants to know how the way things seen from the "first-person perspective" match with data from the "third-person perspective" i. e. from other sources such as developmental psychology and cognitive neuroscience. In psychoanalytic empirical research the psychoanalyst as psychoanalytic researcher is able to integrate the special knowledge of psychoanalysis into research designs.

According to Karl Pribram the recent resurgence of interest in biology has reached also psychoanalysis. Now, therefore more than ever, Freud's "Project for a scientific psychology" (Freud 1950a [1895]) has relevance for those psychoanalysts and other mind-scientists working within the general framework provided by the psychoanalytic metapsychology. In his contribution "Freud's *Project for a Scientific Psychology* in the 21st Century" Pribram summarizes some of Freud's propositions regarding brain function in the light of what we have learned during the past century. Pribram's interest in Freud's "Project" was aroused by two propositions: first, a memory based theory of motivation and second, an Ohm's law of neural processing. Two further areas of interest were developed by Pribram as he and Merton Gill began their studies prior to the publication of their book *Freud's Project Reassessed* (Pribram & Gill 1976): the brain mechanisms involved in primary and secondary processes as well as the brain mechanisms in unconscious and especially conscious processes.

In her contribution Regina Pally "Emotion and the role of emotion in mental life" explains that emotions are not just subjective feelings. Emotions are as much about the body as are digestion and respiration. What neuroscience indicates is that these bodily aspects of emotion are very relevant to the clinical issues we address in psychoanalysis. They help to explain many of the symptoms of psychological disorders, and why they are so deeply entrenched. They also help to explain how empathy, transference, countertransference and psychological defenses occur, and even address the patient's often asked question "why do I have to feel my feelings?" Although emotion is now center stage in neuroscience and in psychoanalysis, there is one caveat. Some emotions are so deeply engrained we cannot change them, even with the best analysis. These patients will continue to have the old painful feelings and will have the impulse or tendency to act on those feelings. The experienced analyst Pally tell patients we have no control over how we feel, only over how we act. This need not be a negative outcome however, because patients in analysis, who are able to bring their emotions into conscious awareness, can learn to inhibit the maladaptive behavior, which results from that emotion.

Fred Levin provides a sample of neuropsychoanalysis. Levin is a pioneer in this field (Levin 1991, 2004), who combines psychoanalytic data with neuroscientific data, the "first-person perspective" and the "third-person perspective". Levin's "Synapses, cytokines and long-term memory. An interdisciplinary look at how psychoanalysis activates learning via its effects on emotional attention" is a paper about the novel neural learning networks and about what happens in mind/brain in relation to psychoanalysis that facilitates learning. When Levin began investigating how learning works, he discovered in 1980 a close link between metaphors coded into analytic interpretations and "aha" reactions in the patient. Levin speculated that such moments of insight were secondary to the simultaneous activation of the primary sensory association cortical zones for hearing, touch, and vision (i.e. the formation of novel neural networks). Later Levin learned from research by Neils A. Lassen suggesting that when an analytic patient feels relaxed and safe he or she becomes more spontaneous as seen in the appearance of free association, transference, and enactments of various kinds. What is important is that in association with such spontaneity, working memory is activated for what is remembered, and this is what makes the patient's learning much easier than usual. After a further review, Levin has recognized that it is possible to apply what is known about synapses, cytokines, and long-term memory, to appreciate that the neural network formation Levin and others speculated about in earlier research would likely be the result of cytokine effects on synaptic plasticity secondary to emotional attention. Insights are first associated with novel neural networks, long-term potentiation (LTP) or long-term depression (LTD), and eventually with new long-term memory (LTM). Learning in analysis thus involves three related activity levels: first, analytic spontaneity (free association and transference being the primary elements), along with, second, emotional attention, which reflects both superficial and deep affective motivational systems of the brain, and third, changes in the chemical matrix which the mind begins to work upon during any analysis. Regarding this third level of activity, which is the primary focus of Levin's paper, the analyst/patient couple facilitates the patient's creation of a state in which new synapses form on demand, simultaneously creating a spatially dispersed collection of neurons, leading to the formation of a novel neural network with the properties. Levin (1991) long ago elaborated on this, in which memory is highly activated and available to the patient (especially working memory for the specific details recalled), and intersensory bridging invites comparisons (based on similarity and difference analysis) and novel insights. Consequently, neuropsychoanalysts can begin to correlate what the analyst does that facilitates a psychological cascade of affective attention, spontaneity, free association, transference, recollection and insight, and a parallel chemical cascade involving the patient's changing synaptic plasticity within systems under the control largely of cytokines. Regarding the psychological and the chemical cascade, at present it is not plausible that one cascade is primary and the other secondary; rather it seems more likely that these two cascades represent the recurrent interactive effects of mind on brain and brain on mind. When we are strongly moved emotionally, we produce changes in synaptic plasticity that further enable us to be so moved.

5. EMPIRICAL RESEARCH TRENDS IN PSYCHOANALYSIS

In addition to the tradition of "psychotherapy research", that originated mainly as outcome-oriented research, "psychoanalytic" research on outcome and process not only learned a lot from psychotherapy research, but it developed several of its own instruments as well as its own strategies (for instance: whereas psychotherapy is usually a short-term treatment, psychoanalytic treatment is a high-frequency and long-term treatment) (Shapiro & Emde 1995). The psychoanalytic assessment instruments try to convert authentic psychoanalytic knowledge into reliable measurement techniques. For instance the Object Relations Inventory was created by Sidney Blatt and colleagues (Blatt, Wein, Chevron & Quinlan 1979; Blatt, Chevron, Quinlan, Schaffer & Wein 1988; Blatt, Levy & Shaver 1998) to assess the structure and the contents of the self and the object representations of clinical subjects, mostly in order to measure the level of psychopathology, but also of non-clinical subjects. In developing the Object Relations Inventory and coding systems, Blatt and colleagues attempted to investigate how the contents and the structure of mental representations are involved in normal development, pathology and therapeutic change. The Object Relations Inventory is administered as an interview measure (although it can also be used as a self-report instrument). Blatt and colleagues developed two scales: First, "Qualitative and Structural Dimensions of Parental Descriptions" and second, "Differentiation-Relatedness Scale of Self and Object Representations" (Diamond, Blatt, Stayner & Kaslow 1991).

The latter scale can integrate clinical observations about early processes of separation-individuation, the formation of the sense of self and the development of interpersonal relatedness. While the Reflective Functioning Scale, which I already mentioned above, emphasizes some cognitive aspects of the representations of self and other, the Differentiation-Relatedness Scale points out more the affective and relational aspects of these representations.

Developmental research in psychoanalysis applies observation techniques to children and to adolescents such as Stuart Hauser's ground breaking empirical research on adolescents "Overcoming adversity in adolescence: narratives of resilience". According to Hauser resilience refers to unforeseen adaptation in the face of serious adversity. Through his narrative analyses of resilient young adults, Hauser has become aware of experiences of self and of relationships that had not been found through previous empirical procedures. For instance, more traditional analyses were theory-driven and constrained by rigorous coding conventions and technical language as the researchers pursued indices of defenses, adaptive strengths, self-images, expressed effects, and enabling interactions. In working with previous methods, Hauser often had the impression that he was *not* capturing significant aspects of the subject's experience; especially their vividly described poignant past and current relationships. Based on these explorations, Hauser concludes that there are good reasons to add narrative analyses to his ongoing life span studies. From Hauser's examination of more than 60 adolescent and adult interviews from the resilient patients – spanning over 20 years – he has a better grasp of resilient participants' constructions of themselves and their relationships during a time of major disruption; and he has seen ways these constructions change over time. In terms of the constructions of the *self*, Hauser found five content themes and one structural feature. First, *self-reflection*, this is illustrated by individuals' increasing awareness of their feelings and thoughts, within each adolescent year and in subsequently reflecting on their experience and performance as adults and new parents. Second, in the case of *agency* as adolescents, the resilient patients played an active part in deciding where they would go after leaving hospital and how they would take care of themselves when facing adversity again. As adults, they make conscious choices about parenting and how to apply their ideas in practice. Relating to this theme, the resilient adults' detailed visions of their future, and their conscious plans enable them to actualize these optimistic aspirations. Third, indicative of *self-complexity*, Hauser found evidence of complex continuities over time, the resilient subjects' recognizing parenting themes from their past; they accept their discovery of multiple perspectives about themselves, and about their friends. Fourth, regarding *persistence and ambition*, resilient patients describe a refusal to settle for a specific solution offered by a hospital, a therapist, or their family; several find new schools or return to complete training many years after forced breaks. Fifth, concerning *self-esteem,* ever apparent are the resilient participants' vacillating appraisals of themselves. By no means were these evaluations simply increasingly positive over the years. More significantly was their awareness of these self-evaluations, and the overall balance weighted in favor of more positive self-regard each year. Concerning the structural feature these were *coherent narratives*. Focusing on constructions of *relationships*, Hauser identifies three themes: the resilient adolescents strongly reflect on others' motives, feelings, and thoughts. Second, the resilient individuals attribute great importance to friendships. Finally, the resilient individuals are aware of many *intersections* between themselves, their relationships, and their actions; speaking, for instance, about how increasingly good feelings about themselves are leading to more positive ways in which they are searching for new friends.

As part of a wider attempt to understand and investigate at an empirical level, what we experience at a clinical level, i. e. a deep, stable change in our patients, the psychoanalyst and psychoanalytic researcher Rolf Sandell, director of the Stockholm Outcome of Psychoanalysis and Psychotherapy Project (STOPP-Project), has considered the definition and assessment of "structural change" *a posteriori* and not directly. In his study Sandell examines and compares the effects of psychoanalysis and long-term-therapy. From the point of view of the clinicians it remains the need to put together such impressive empirical findings with their own clinical experience. In his contribution for this volume "Structural change and its assessment. Experiences from the Stockholm Outcome of Psychoanalysis and Psychotherapy Project" Sandell analyzes the favoured psychoanalytic clinical concept of structural change with some questions in mind. This includes: What are the essential aspects of the kind of change that may be called structural (in contradistinction from non-structural kinds of change)? Is the structural kind of change specific to psychoanalysis or may it come about in other ways? Are all changes during psychoanalysis, structural? How may structural change be properly assessed? In his *Introductory Lectures on Psycho-Analysis* Freud asked himself if psychoanalysis was to be defined as a "causal" therapy opposed to symptomatic therapies, which do not search for the causes but treat only the effects of the illness (Freud 1916–17 a [1915–17], pp. 435–436). In the original STOPP-Project Sandell did not integrate a particular investigation or measure for assessing structural change *a priori*. This investigation occurred only later on. This is because at the beginning Sandell assumed a neutral point of view. He did not take for granted that structural change occurs in psychoanalysis and even questioned that it has to be considered a peculiarity of psychoanalysis. Nevertheless according to Sandell one of the important results of the STOPP-Project is that psychoanalysis *may be* expected to produce stable change, i. e., structural change. Sandell differentiates between three kinds of "structural change": first, structuring or stabilizing a system that is less stable, for example an increased capacity of the ego to bind energy; second, restructuring, for instance a non-adaptive solution to a conflict is replaced by an adaptive solution, the super-ego becomes more benign; third destructuring or destabilizing, i. e. in the case of a patient suffering from an obsessive-compulsive neurosis. According to Sandell a "structural change" is a permanent, enduring change, which is unobservable and has to be inferred on the basis of indicators of underlying structures, when symptoms disappear, and no new symptom emerges, this indicates structural change. Referring to his own definition of structural change, this can only be indirectly assessed. Sandell uses instruments such as self-rating personality inventories, observer rating scales, and other measures; the important difference is that in his research design Sandell makes repeated observations, *exploring* the consistency of the phenomenon of a stable, permanent, enduring "structural" change rather than just expecting it. Assuming that stabilization does not depend on the clinician or on the instrument being insensitive to change, or on the possibility that the patients respond to the basis of response sets such as social desirability, in order to please the analyst or therapist, Sandell was able to draw more reliable conclusions about what amount and kind of structural change had taken place.

In "On the contribution of the most recent research on language applied to psychoanalytic practice", the psychoanalyst Jorge Canestri links current findings of linguistics to psychoanalytic clinical practice. He presents some ideas on the use of language in psychoanalysis in the light of what psychoanalysis has learned from linguistics, neurolinguistics and the neurosciences in general during recent years. From a linguistic perspective, Canestri's starting point is to consider, following Chomsky, that the knowledge of language falls within the mind/brain human system (I-language). This makes linguistics a "science of nature" and opens the field to interdisciplinarity and, consequently, to a study of the elements that, in maturational stages and with specific modalities, contribute to the development of the child's "linguistic system". This system must be considered in close relation to the psychological development in order to analyze the quality of the representations and of the symbolizations that will inform the future mental life of the subject. Through the use of recently developed techniques of neuroimaging, neuroscience gives psychoanalysis the opportunity to revalue some of its own theories about the localizations of the linguistic functions. However even more interesting is the view that they offer us first the opportunity to investigate the acquisition of one's mother tongue in the light of the distinction between procedural and declarative memory, and second of differentiating, in terms of development, between the precocity of the auditory system and that of recognition and the relative delay of the phonic system. This is associated to precocity in understanding the meaning of communications which is certainly superior to children's effective linguistic realizations and actual lexical capacities. Canestri proposes that, when referring to the "linguistic system", we regard it as a complex semiotic system, inasmuch as in the case of infantile linguistic development, linguistic communication is inseparable from gestic and postural expression and from the use of multiple channels of communication, such as physical contact, glances, etc. The study of mother/environment-child communication contributes, also in the adult age, to a better comprehension of the structure underlying the therapeutic interaction. In fact, an actual theory on linguistic acquisition and on the capacity of symbolization should be based on pragmatic criteria, i.e. on the real levels of affective and linguistic exchange and on the reciprocal affective monitoring between mother/environment-child. Furthermore, it should be taken into consideration that a large part of what takes place in the relationship between patient and psychoanalyst, in terms of prototypical reactions which repeat models of relationship with the primary objects, can be related to procedural or implicit memory. These two types of memory consist of two different registers of phenomena, i.e. multiple parallel channels for processing information that use different codes and modalities. Psychoanalysts being aware of these two types of memory are faced with the problem of continuously carrying out processes of translation, or of transformation, that link the incoming information to a common representation that can only be symbolic, that has to be shared with the patient, and that has to make use of words. The application of this knowledge to the psychoanalytic field leads Canestri to suggest an update of some of the criteria that up until now have oriented therapeutic action, and to propose a paradigm alternative to the classical one.

REFERENCES

Agassi, J. (1975). *Science in Flux*. Dordrecht: Reidel.

Baron-Cohen, S. (1995). *Mindblindness: En Essay on Autism and Theory of Mind.* Cambridge, Mass.: MIT Press.

Baron-Cohen, S., Tager-Flusberg, H. & Cohen, D.J. (1993). *Understanding Other Minds: Perspectives from Autism.* Oxford: Oxford University Press.

Bieri, P., ed. (1993). *Analytische Philosophie des Geistes*. Bodenheim: Athenäum.

Blatt, S.J., Chevron, E.S., Quinlan, D.M., Schaffer, C.E., & Wein, S.J. (1988). The assessment of qualitative and structural dimensions of object representations (revised edition). Unpublished research manual. New Haven, Yale University.

Blatt, S.J., D'Afflitti, J. P., & Quinlan, D. M. (1976). Experiences of depression in normal young adults. *Journal of Abnormal Psychology* 85: 383–389.

Blatt, S.J., Levy, K.N., & Shaver, P.R. (1998). Attachment styles and parental representations. *Journal of Personality and Social Psychology* 74: 407–419.

Blatt, S.J., Wein, S.J., Chevron, E., & Quinlan, D.M. (1979). Parental representations and depression in normal young adults. *Journal of Abnormal Psychology* 88: 388–397.

Block, N. (1995). On a confusion about a function of consciousness. *The Behavioral and Brain Sciences*, 18: 227–247; also in *The Nature of Consciousness. Philosophical Debates*, ed. N. Block, O. Flanagan. & G. Güzeldere. Cambridge, Mass.: MIT Press 1997, pp. 375–415.

Block, N., Flanagan, O. & Güzeldere, G., ed. (1997). *The Nature of Consciousness. Philosophical Debates*. Cambridge, Mass.: MIT Press.

Bott-Spillius, E. (1992). Clinical experiences of projective identification. In *Clinical Lectures on Klein and Bion*, ed. R. Anderson. London, New York: Routledge, pp. 59–73.

Brentano, F. (1874). *Psychology from an Empirical Standpoint.* Vol. 1. New Jersey: Humanities Press 1973.

Brentano, F. (1924). *Über die Zukunft der Philosophie*. Hamburg: Meiner 1968.

Chalmers, D. (1996). *The Conscious Mind.* Oxford: Oxford University Press.

Davidson, D. (1970). Mental events. In *Essays on Actions and Events*. Oxford: Clarendon Press 1980, pp. 207–225.

Davidson, D. (1984). First person authority. *Subjective, Intersubjective, Objective*. Oxford: Clarendon Press 2001, pp. 3–14.

Davidson, D. (1987). Knowing one's own mind. *Subjective, Intersubjective, Objective*. Oxford: Clarendon Press 2001, pp. 15–38.

Davidson, D. (1993). Reply to Peter Bieri. In *Reflecting Davidson*, ed. R. Stoecker. Berlin & New York: De Gruyter, pp. 311–313.

Diamond, D., Blatt, S. J., Stayner, D.A. & Kaslow, N. (1991). Self-other differentiation of object representations. Unpublished research manual. New Haven, Yale University.

Dretske, F. (1995). *Naturalizing the Mind.* Cambridge: MIT Press.

Farah, M.J. (1994). Visual perception and visual awareness after brain damage: a tutorial overview. In *Attention and Performance*, ed. C. Umiltà & M. Moscovitch. Cambridge, Mass.: MIT Press, pp. 37–75.

Farrell, B.A. (1950). Experience. *Mind* 59: 170–198.

Fisher, S. & Greenberg, R.P. (1985). *The Scientific Credibility of Freud's Theories and Therapy*. New York: Columbia University Press.

Fisher, S. & Greenberg, R.P. (1996). *Freud Scientifically Reappraised: Testing the Theories and Therapy*. New York: Wiley.

Fisher, S. & Greenberg, R.P. (2002). Scientific tests of Freud's theories and therapy. In *The Freud Encyclopedia. Theory, Therapy and Culture*, ed. E. Erwin. New York, London: Routledge, pp. 509–514.

Flanagan, O. (1992). *Consciousness Reconsidered.* Cambridge, Mass: MIT Press.

Fodor, J. (1975). Special sciences, or the disunity of science as a working hypothesis. In *The Language of the Thought*. New York: Crowell, pp. 9–25.

Fonagy, P., Gergely, G., Jurist, E.L. & Target, M. (2002). *Affect Regulation, Mentalization, and the Development of the Self.* New York: Other Press.

Fonagy, P., Target, M., Steele, H., & Steele, M. (1998). Reflective-Functioning manual. 5th edition. Unpublished research manual. London, University College London.

Foucault, M. (1967). Nietzsche, Freud, Marx. In *Nietzsche. Cahiers de Royaumont.* Paris: Minuit, pp. 183–192.

Freud, S. (1890a). Psychical (or mental) treatment. *S.E.* 7: 281–302.

Freud, S. (1891b) (1). *On Aphasia.* New York: International Universities Press 1953.

Freud, S. (1891b) (2). *Zur Auffassung der Aphasien*, ed. P. Vogel. Frankfurt am Main: Fischer 1992.

Freud, S. (1900a). *The Interpretation of Dreams*. *S.E.* 4.

Freud, S. (1905d). *Three Essays on Sexuality. S.E.* 7: 123–243

Freud, S. (1915e). The unconscious. *S. E.* 14: 159–207.

Freud, S. (1916–17a [1915–17]). *Introductory Lectures on Psycho-Analysis. S.E.* 16.

Freud, S. (1925d [1924]). *An Autobiographical Study. S.E.* 20: 7–70.

Freud, S. (1925e [1924]). The resistances to psycho-analysis. *S.E.* 19: 213–222.

Freud, S. (1926e). *The Question of Lay Analysis. S.E.* 20: 177–258.

Freud, S. (1933a [1932]). *New Introductory Lectures on Psycho-Analysis. S.E.* 22.

Freud, S.(1940a [1938]). *An Outline of Psychoanalysis. S.E.* 23: 139–207.

Freud, S. (1940b [1938]). Some elementary lessons in psycho-analysis. *S.E.* 23: 281–286.

Freud, S. (1950c [1895]). Project for a scientific psychology. *S.E.* 1: 295–397.

Freud, S. (1955e [1930]). Brief an Juliette Boutonier. *Nachtragsband. Gesammelte Werke.* Frankfurt am Main: Fischer 1987, pp. 671–672.

Freud, S. (1985). *The Complete Letters of Sigmund Freud to Wilhelm Fließ 1887–1904*, ed. J. M. Masson. Cambridge, Mass.: Harvard University Press.

Freud, S. & Ferenczi, S. (1993). *The Correspondence of Sigmund Freud and Sándor Ferenczi 1908–1914.* Vol. 1, ed. E. Brabant, E. Falzeder, & P. Giampieri-Deutsch. Cambridge, Mass.: Harvard University Press.

George, C., Kaplan, N. & Main, M. (1996). Adult Attachment Interview. 3rd edition. Unpublished research manual. Berkeley, Department of Psychology, University of California.

Giampieri-Deutsch, P. (1996). The influence of Ferenczi's ideas on contemporary standard technique. In *Ferenczi's Turn in Psychoanalysis*, ed. P. Rudnytsky, A. Bókay & P. Giampieri-Deutsch. New York: New York University Press, pp. 224–247.

Giampieri-Deutsch, P. (1997). Die Psychoanalyse als Wissenschaft und andere Aspekte des Briefwechsels zwischen Sigmund Freud und Sándor Ferenczi. In *Jenseits von Kunst*, ed. P. Weibel. Wien: Passagen Verlag, pp. 621–626.

Giampieri-Deutsch, P. (1999). Freud und der gegenwärtige Stand der philosophischen Untersuchungen zur Theorie des Mentalen. *Wiener Jahrbuch für Philosophie* 31: 229–256.

Giampieri-Deutsch, P., ed. (2003a). *Psychoanalyse im Dialog der Wissenschaften. Europäische Perspektiven.* Vol. 1. Stuttgart: Kohlhammer.

Giampieri-Deutsch, P. (2003b). Die psychoanalytische Theorie des Mentalen und die analytische Philosophie des Geistes. In *Psychoanalyse im Dialog der Wissenschaften. Europäische Perspektiven.* Vol. 1, ed. P. Giampieri-Deutsch. Stuttgart: Kohlhammer, pp. 58–75.

Giampieri-Deutsch, P., ed. (2004a). *Psychoanalyse im Dialog der Wissenschaften. Amerikanische Perspektiven.* Vol. 2. Stuttgart: Kohlhammer.

Giampieri-Deutsch, P., ed. (2004b). Auswirkungen der Kooperation zwischen der analytischen Philosophie des Geistes und der Psychoanalyse auf die empirische psychoanalytische Forschung. *Psychoanalyse im Dialog der Wissenschaften. Amerikanische Perspektiven.* Vol. 2, ed. P. Giampieri-Deutsch. Stuttgart: Kohlhammer, pp. 80–91.

Grinberg, L. (1979). Countertransference and projective counteridentification. In *Countertransference*, ed. L. Epstein & A.H. Feiner. New York & London: Jason Aronson, pp. 169–191.

Grünbaum, A. (1984). *The Foundations of Psychoanalysis.* Berkeley: University of California Press.

Grünbaum, A. (1993). *Validation in the Clinical Theory of Psychoanalysis. A Study in the Philosophy of Psychoanalysis. Psychological Issues Monograph* 61. Madison: International Universities Press.

Grünbaum, A. (2002). Critique of psychoanalysis. In *The Freud Encyclopedia. Theory, Therapy and Culture*, ed. E. Erwin. New York, London: Routledge, pp. 117–136.

Grünbaum, A., ed. (1991). *Kritische Betrachtungen zur Psychoanalyse. Adolf Grünbaums „Grundlagen" in der Diskussion.* Berlin, New York: Springer.

Güzeldere, G. (1997). Introduction. The many faces of consciousness: a field guide. In *The Nature of Consciousness. Philosophical Debates*, ed. N. Block, O. Flanagan & G. Güzeldere. Cambridge, Mass.: MIT Press, pp. 1–67.

Güzeldere, G., Flanagan, O. & Hardcastle, V.G. (2000). The nature and function of consciousness: Lessons from blindsight. In *The New Cognitive Neurosciences*, ed. M. Gazzaniga. Cambridge, Mass.: MIT Press, pp. 1277–1284.

Habermas, J. (1968). *Knowledge and Human Interest*. Boston: Bacon Press 1971.

Jackson, F. (1982). Epiphenomenal qualia. *American Philosophical Quarterly* 32: 127–136.

Jackson, F. (1986). What Mary didn't know. *Journal of Philosophy* 83: 291–295

Klein, M. (1946). Notes on some schizoid mechanism. *International Journal of Psycho-Analysis* 27: 99–110.

Levin, F.M. (1991). *Mapping the Mind. The Intersection of Psychoanalysis and Neuroscience*. Hillsdale, London: The Analytic Press.

Levin, F.M. (2004). *Psyche and Brain. The Biology of Talking Cures*. Madison: International Universities Press.

Milner, B., Squire, L.R., & Kandel, E.R. (1998). Cognitive neuroscience and the study of memory. *Neuron* 20: 445–468.

Nagel, E. (1959). Methodological issues in psychoanalytic theory. In *Psychoanalysis, Scientific Method and Philosophy*, ed. S. Hook. New York: New York University Press, pp. 38–56.

Nagel, T. (1974). What is it like to be a bat? In *The Nature of Consciousness. Philosophical Debates*, ed. N. Block, O. Flanagan & G. Güzeldere. Cambridge, Mass.: MIT Press 1997, pp. 519–527.

Ogden, T.H. (1982). The analytic management and interpretation of projective identification. In *Countertransference. Theory, Technique, Teaching.* London: Karnac Book, pp. 21–46.

Popper, K. (1962). *Conjectures and Refutations*. New York: Basic Books

Pribram, K.H., & Gill M. (1976). *Freud's Project Re-assessed*. New York: Basic Books.

Ricœur, P. (1965). *De l'interprétation. Essai sur Freud.* Paris: Editions du Seuil.

Rubinstein, B.B. (1967). Explanation and mere description: A metascientific examination of certain aspects of the psychoanalytic theory of motivation. In *Motives and Thought. Psychoanalytic Essays in Honor of David Rapaport. Psychological Issues Monograph* 18–19, ed. R. Holt. New York: International Universities Press, pp. 18–78.

Schlick, M. (1925). *Allgemeine Erkenntnislehre*. Frankfurt am Main: Suhrkamp 1979.

Searle, J. (1992). *The Rediscovery of the Mind.* Cambridge, Mass.: MIT Press.

Shapiro, T. & Emde R.N., ed. (1995). *Research in Psychoanalysis: Process, Development, Outcome*. Madison Connecticut: International Universities Press.

Shoemaker, S. (1993). The first-person perspective. In *The Nature of Consciousness. Philosophical Debates*, ed. N. Block, O. Flanagan & G. Güzeldere. Cambridge, Mass.: MIT Press 1997, pp. 503–515.

Squire, L.R. & Kandel E.R. (1999). *Memory. From Mind to Molecules*. New York: Scientific American Library.

Van Gulick, R. (1993). Understanding the phenomenal mind: are we all just armadillos? In *Consciousness: Psychological and Philosophical Essays*, ed. M. Davies & G. Humphreys. Oxford: Blackwell, pp. 137–154.

Velmans, M., ed. (2000). *Investigating Phenomenal Consciousness*. Amsterdam, Philadelphia: John Benjamins Publishing Company.

Waelder, R. (1962). Psychoanalysis, Scientific Method, and Philosophy. *Journal of the American Psychoanalytic Association* 10: 617–637.

Wittgenstein, L. (1953). *Philosophical Investigations*, ed. G.E.M. Anscombe, G.H. v. Wright & R. Rhees. Oxford: Blackwell.

A. Psychoanalysis, Philosophy of Mind and Experimental Sciences: Research on Conscious and Unconscious Processes

II. TOWARD A THEORY ON CONSCIOUSNESS BASED ON RECENT DEVELOPMENTS IN SUBLIMINAL RESEARCH

HOWARD SHEVRIN

1. INTRODUCTION

In this paper I will be presenting the thesis that one of the most important questions in contemporary psychology and neuroscience concerns the relationship between conscious and unconscious processes. It will be my contention that consciousness cannot be fully understood without a concomitant understanding of unconscious processes, and that unconscious processes cannot be fully fathomed without an understanding of conscious processes. It will be my further contention that this conjoint understanding can only be achieved by placing individuality at the heart of the solution. It is widely accepted by consciousness theorists that subjectivity is the hallmark of consciousness. Subjectivity by definition is always individual. From the standpoint to be developed in this contribution, there cannot be a completely free standing theory of consciousness without taking the unconscious into account, or a theory of the interaction between conscious and unconscious processes without taking individual differences into account. It follows from this position that any quantitative or qualitative differences between conscious and unconscious processes can best be seen through the lens of individual differences. Methods must be devised to study conscious and unconscious processes jointly as a function of these individual differences. I will be describing a number of studies that illustrate this position.

Until recently the interaction of conscious and unconscious processes would not have been of interest to psychologists and neuroscientists. For half of the twentieth century psychology was denying the relevance of consciousness to a scientific psychology, and for several more decades it disputed the existence of unconscious processes. A decade before the turn of the twentieth century, James published ten arguments against the concept of an unconscious, the last argument leveled at the existence of unconscious motivation (James 1890). Not long afterwards, Watson ruled consciousness out of bounds for scientific investigation. In an ironic footnote to history, Watson's famous Alfred experiment was intended to provide experimental evidence for the psychoanalytic theory of childhood phobias. To my knowledge this experiment was the high water mark of Watson's flirtation with psychoanalysis. Following Watson, Skinner emerged as the foremost figure in psychology. For Skinner the notion of a mental life, conscious or unconscious, was subject matter for literature and not for science. He also rejected the relevance of neuroscience to psychology. Of the leading figures of twentieth century psychology it was only Freud who upheld the importance of a mental life grounded in the brain and characterized by interacting conscious and unconscious processes in which feeling and desiring in all its forms occupied an important place alongside perceiving, remembering, and thinking. Equally important to Freud was the importance of personal history and the unique contexts in which individual experience occurs; psychoanalytic theory in large part emerged out of clinical experience with individual patients. In this regard it is of interest that only recently have models of brain functioning begun to appear which posit highly interactive neuronal systems in which individual differences, context and history are essential factors (Edelman 1987; Friston 2000). Similarly, in an official report of the American Psychological Society, the home of most American cognitive scientists, Brewer and Luce (1998) strongly recommended that future cognitive research incorporate individual differences and context.

As we begin the twenty-first century the challenge to psychology and neuroscience is to create vibrant, real to life explanations of how the brain/mind organization works. The programmatic Freudian perspective on the mind in its broadest essentials can be of assistance in this task. As Kandel (1999) has observed with respect to the current status of psychiatry that could apply as easily to psychology and neuroscience, psychoanalysis offers, for all its limitations, the only comprehensive theory we have. At the same time, psychoanalysts must see beyond the narrow boundaries of their clinical enterprise to the broader horizons envisioned by Freud's comprehensive theory of mind, and embrace the necessity for empirical and experimental methods, along with greater rigor of systematic thinking.

To better capture the essence of what I wish to convey I will start with a metaphor. Imagine a stage on which a play is being produced, in some ways different from Dr. Baars' *Theater of Consciousness* (1997). The actors deliver their lines, interact with each other, and then exit singly or severally, as the plot unfolds. In the ordinary theater the actors exit into the wings to have a cup of coffee, change costumes, or they rehearse their next lines and otherwise prepare themselves for their return on stage. But in my theater they are never off stage. When they exit one stage they enter on to a new stage hidden from the audience and the other actors where they continue to interact with each other. The audience in front of the visible stage is unaware of this other stage, while the actors you see come and go from one stage to the other are only partially aware of what has been happening on the invisible stage. Yet what is happening on it will greatly determine what happens next in the play. When the actors now reenter from the hidden stage, sometimes on cue, sometimes not, they are saying things and reacting to the other characters in ways that are puzzling and disconcerting to the other actors and to the audience. But being seasoned actors and theatergoers they try to paper over these disconnections and disruptions and maintain some sense of continuity. Nevertheless there is much that is happening on stage that is mystifying and at times disorganized, replete with elisions, omissions, and seemingly out of place speeches and actions, a kind of theater of the absurd.

I first presented this analogy in discussion remarks at a CIBA conference on consciousness (Bock & Marsch 1993). To my surprise, some years later the British playwright, Alan Ayckbourn wrote two plays with the same actors to be performed simultaneously on two stages, one play taking place indoors and the other in an adjacent garden outdoors. The plays were called *Home and Garden* (2000). When the actors exited from one stage it was to enter the second stage, and what happened before and after on either stage influenced the action seen by the two separate audiences. Playgoers were encouraged to see both plays so that they would have a complete appreciation of what was going on separately on each stage. In my analogy the one play was the unconscious for the other.

There are several points I wish to make by way of this theater analogy:

1. Consciousness cannot be fully understood without knowledge of what is going on unconsciously. The audience in my metaphor cannot finally understand what is going on before them without knowing about the events occurring on the other invisible stage.

2. What is going on unconsciously is as complex and representational as what is going on consciously. It is not the case that unconscious processes are essentially part processes that are only integrated consciously, or that they are purely physiological and it is only with consciousness that the truly mental in Brentano's sense of possessing intentionality appears, nor is the unconscious purely dispositional or latent. The actors on my invisible stage are as real as the actors on the

visible stage, in fact they are the same actors, their lines are as meaningful, and their actions are as consequential.

3. The interactions between conscious and unconscious processes are always taking place at a specific time, in a particular place, and in one person only whose history invests with particularity the form and content of the interaction. Thus the interaction cannot be understood apart from context. My play on two stages deals with particular events in the lives of particular people. The play itself is the expression of one mind and the product of one brain.

4. Governing the interaction between conscious and unconscious processes are certain regularities that emerge out of the individual instantiation of these interacting processes. These differences are of a quantitative and qualitative nature, as I will try to illustrate from our research.

Mainly I will be concentrating my remarks on quantitative and qualitative differences in the interaction of conscious and unconscious processes because these differences will allow me to illustrate the other points just described as well. In this short chapter I will describe these views in schematic form, although I will refer to former published reports or research in progress that I and my colleagues have conducted in support these views. Among these colleagues are Drs. Linda Brakel, Michael Snodgrass, William Williams, Ramesh Kushwaha, Selin Aviyente, Edward Bernat, Scott Bunce, Philip Wong, Karen Villa, Shasha Kleinsorge, John Hartman, Richard Hertel, and James Bond. Our group is interdisciplinary, made up of psychologists, psychoanalysts, psychiatrists, and biosignal analysts.

2. QUANTITATIVE RELATIONSHIPS BETWEEN CONSCIOUS AND UNCONSCIOUS PROCESSES

To return to the theatrical analogy, and if I may strain it to the breaking point, the reason the "unconscious" stage remains unknown to the audience is that the action on it is carried out in whispers. In short, there is a quantitative factor that is the best researched property of unconscious processes, at least for those unconscious processes investigated with subliminal stimuli. Stimulus energy has a good deal to do with whether a stimulus will become conscious or not. Yet this is not the whole story. Our research has shown that there is a complex relationship between stimulus energy and different kinds of thresholds that have quite different implications for the relationship of the stimulus to consciousness.

By way of introduction to this point let me show you a plot of event-related potentials to the same stimuli presented subliminally and supraliminally. At first glance the subliminal curve appears to be a flat line as compared to the noticeable amplitude fluctuations in the supraliminal curve. However, if you simply multiplied each point on the subliminal curve by a factor of four, the resultant curve is markedly similar in structure to the supraliminal curve. In fact, the cross-correlations at a number of electrodes are quite substantial and highly significant statistically. In other words, the processes going on unconsciously are similar to those going on consciously; they differ solely in their strength, or in this case, in their voltage amplitude. (This plot was published in a paper by Bernat, Bunce and Shevrin in the *International Journal of Psychophysiology* in 2001 as part of a study of unconscious affect).

But the full story about stimulus intensity is more complicated and interesting. In a searching methodological and theoretical examination of subliminal research Dr. Snodgrass in collaboration with Dr. Bernat and myself in a paper to be published in *Perception and Psychophysics* (2005) concluded that much recent subliminal research is conducted at the subjective threshold. At this threshold phenomenal consciousness is present despite the subject's avowal that only guessing is involved. We distinguish between phenomenal and reflective consciousness, the latter involving an awareness of being aware whereas the former does not. This is similar to the distinction made by Edelman (1987) between primary and secondary consciousness, and Block (1995) between phenomenal consciousness and access consciousness. At the subjective threshold the stimulus conditions fail to meet the criteria for reflective awareness but phenomenal awareness is present. The stimuli are thus not truly unconscious.

Another way of describing this distinction can be drawn from signal detection theory. At the subjective threshold d', or stimulus detectability, is significantly above zero, but the avowed absence of consciousness is a function of the subject's criterion for what is deemed sufficiently clear to be accepted as conscious. Or otherwise put, the subject reflects on, or judges whether a given impression in phenomenal consciousness meets the subject's personal criterion for consciousness. By contrast, there is a body of subliminal research, including our own studies that have been conducted below the objective detection threshold. At the objective detection threshold d' is at zero, no phenomenal consciousness of the stimulus is present.

Of great relevance to the relationship of conscious and unconscious processes is the finding that a different relationship exists between stimulus detection, d', and the purported subliminal effects at the subjective and objective thresholds. (In most subliminal studies the test for subliminality is administered *after* the assessment of a subliminal effect on some task has been obtained). At the subjective threshold a *positive* relationship exists between d' and the purported stimulus effect: The more sensitive the subject is to the stimulus on the detectability task, the greater the subliminal effect. This is exactly what one would expect if the stimulus were in phenomenal consciousness. It is quite otherwise at the objective detection threshold. The relationship is *negative* (Bernat, Shevrin & Snodgrass 2001; Snodgrass, Shevrin & Kopka 1993). The greater the sensitivity to the stimulus, the *weaker* the subliminal effect. Here there seems to be a curious relationship between a measure of stimulus detectability and whether or not a subliminal effect is strong or weak.

At first glance it might appear that if even a smidgen of consciousness were present this would serve to inhibit subliminal effects. In fact, under certain conditions some consciousness can inhibit subliminal effects. This is particularly true at the objective identification threshold at which subjects can consciously see something but not enough to correctly identify the stimulus as a particular word, for example; subliminal effects tend to disappear under these conditions. Snodgrass (2005b) has developed a nonmonotonic theory to explain these relationships between consciousness and subliminal effects.

But at the objective detection threshold it is difficult to believe that a d' slightly above zero and falling well within a chance distribution around zero would reflect any degree of consciousness. Moreover, statistical analysis reveals that there are no outliers. And yet there must be something systematic in these individual fluctuations above and below the d' mean or we would not have found the negative correlations in more than one study.

The possibility suggests itself that we are dealing with *individually determined* changes in unconscious processing that have different consequences for what will enter consciousness as the response to a subliminal stimulus. In a series of studies we found that when subjects were asked to guess which of four words was flashed subliminally at the objective detection threshold (1 msec; 5 ft/lamb), the main effect was at chance; the proportion of correct guesses was not significantly different from 25%. However, we had also built into the experiments two other factors: 1. task strategy, and 2. task preference. Task strategy took two forms: a "pop" strategy in which subjects were asked to let one of the four words "pop" into mind, and a "look" strategy in which subjects were asked to look as carefully as possible and to base their guesses on what they thought they saw. Subjects were also asked which of the two strategies they preferred. About thirty to forty percent preferred the "look" strategy.

Once strategy and preference were introduced taken into account, rather striking findings emerged: In seven experiments, two conducted in another laboratory (Van Selst & Merikle 1993), we discovered a consistently significant interaction between strategy and preference (Snodgrass, Shevrin & Kopk 1993). Of particular interest were the "look" preference subjects. In the "pop" strategy they performed significantly *below* chance while in the "look" strategy they performed significantly *above* chance. The "pop" preference subjects showed the opposite pattern, but more weakly. Clearly if you were to add up performance in the two strategies you would end up with zero subliminal effect.

Now it could be argued that the "look" subjects became conscious of the stimulus when the strategy fit with their preference, hence they performed above chance in the "look" strategy. If that were the case we should expect to find a *positive* correlation between d' and their correct guesses. It was quite otherwise. The correlation was significantly *negative*. Any individual *increase* in d' worked against improvement in their guesses. In fact, the negative correlation suggests that a *below* chance d' *enhances* guessing performance. And if we bear in mind that unconscious processes are involved whether a score is above or below the d' mean, we can see how there must be *individually determined* fluctuations in unconscious processing that exercise inhibitory or facilitating effects on what enters consciousness.

It is perhaps not accidental that the subjects who show the below chance performance in the "pop" condition are "lookers". These are people to whom seeing what is there is important and who are not disposed to letting things just "pop" into their heads. Since nothing is really there to be seen consciously, they inhibit unconsciously the correct responses. The stake these subjects have in maintaining a coherent conscious experience results in diminishing their unconscious performance. I offer this as an example of how conscious and unconscious processes interact as a function of individuality (strategy preference) and context (task strategy).

In the same series of experiments we also have evidence that defensive operations in the psychoanalytic sense may be involved. The Hysteroid-Obsessoid Questionnaire (HOQ) is a personality inventory devised to identify hysterical patients on the one extreme and obsessive patients on the other extreme (Caine & Hawkins 1963). From other research we have reason to believe that the scale also correlates with a tendency to use repression as a defense (Ludolph 1981). It turns out that subjects high in repression show the same pattern as "lookers". They inhibit in the "pop" condition and facilitate in the "look" condition. Nevertheless, and perhaps surprisingly, "lookers" and "repressors" are not the same people. The two factors are unrelated to each other. The repression finding points to another individual difference playing a role in how subliminal stimuli are processed and what is permitted into consciousness. One could not determine from the below or above chance performance alone whether it was look preference or repression that produced the results.

The relationship between repression and consciousness has been explored in another study recently published by Libet, Ghannam, and myself (2002). This study took as its point of departure Libet's well known finding that it takes on average of 500msec for a stimulus to become conscious (Libet, Alberts, Wright et al. 1964). What is not as well known is that subjects vary from 200 to 800 msec in the time for a stimulus to become conscious. What might be going on during this varying time interval? We reasoned that this individual difference might constitute a necessary condition for repression. People who took a longer time to become conscious would be more likely to use repression as a defense. We tested this hypothesis by administering this same Hysteroid-Obsessive Quesionnaire to Libet's subjects whose time varied from 200 to 750msec. We found as hypothesized that longer time to consciousness was associated with greater repressiveness. This finding suggests that an individual characteristic of consciousness contributes to the selection of a particular defense, another instance of the way conscious and unconscious processes interact as a function of individual differences.

However, there is a problem in too hastily concluding that repression is involved in the studies just described. The psychoanalytic theory of repression requires that the stimuli have *personal* and *unconscious conflictual* significance. The stimuli in neither the pop/look studies nor the time to consciousness study met these three criteria. In another study we undertook to meet these criteria, an initial report of which was published in *Consciousness and Cognition* (Shevrin, Williams, Marshall et al. 1992), and a final book length report some time thereafter (Shevrin, Bond, Brakel et al. 1996). In this study a team of psychoanalysts examined interview transcripts conducted with patients suffering from social phobias. They then arrived at a hypothesis concerning the unconscious conflict causing the social phobia in each patient. Parenthetically, I should note in keeping with the thesis of this contribution that the conflicts were as various as the patients. On the basis of this hypothesis the team of psychoanalysts selected words or brief phrases they believed were significantly related to the unconscious conflict. In addition, words were selected that captured the patient's conscious experience of the phobia, as well as two control categories made up of unrelated negative and positive affect words. These four sets of words were then presented subliminally (below the objective detection threshold) and supraliminally, while at the same time event-related potential brain responses were obtained to each stimulus.

There were two findings of interest: 1. when the stimuli related to the hypothesized unconscious conflict were presented *subliminally* the brain responses differentiated them from the other three sets of stimuli, but when the same unconscious conflict stimuli were presented *supraliminally* the brain responses no longer differentiated them from the other sets of stimuli, 2. the difference between successful differentiaton subliminally and the failure to differentiate supraliminally was positively correlated with the Hysteroid-Obsessiod Questionnaire. The more repressive the patient the greater the difference in favor of subliminal differentiation of the unconscious conflict stimuli. This relationship was not found for the words related to the conscious symptom experience, which as one might expect were better differentiated supraliminally.

Taken together these findings support the inference that repression must have been involved. The words had been selected in advance by a team of psychoanalysts who hypothesized that they were related to the presumed unconscious conflict, these words were highly personal (a point I shall return to), and they were presented out of consciousness, thus meeting the three criteria for repression. And independently supporting this inference was the correlation with the Hysteroid-Obsessoid Questionnaire.

As I just mentioned, the words were highly personal. Unlike most psychological experiments, including some of our own like the pop/look studies, the verbal stimuli used were unique for each subject. This was particularly true for the words related to the unconscious conflict. Nevertheless, the same *functional* relationships were found across subjects. I cite this as an example of how taking individuality into account need not leave us with idiosyncratic ideographic data lacking in generalizability, but rather constitutes the only way that certain general principles can be discovered. In this instance the general principle concerns the operation of defenses with respect to unconscious conflict. But I suggest that when dealing with the interaction between conscious and unconscious processes unless individuality is taken into account important principles will go undiscovered. The time to consciousness discovered by Libet is certainly of general interest, but as our study showed it is remarkably sensitive to individual differences. We have evidence from other studies as well that personality factors are especially important when dealing with unconscious processes as these interact with consciousness. The pop/look studies are good examples of this principle.

The major limitation, however, of the studies just described is that they do not help us to identify the *mechanism* of repression. The time to consciousness study describes a necessary by not sufficient condition for repression. The pop/look and social phobia study provide purely correlational evidence. There is one study currently under way conducted by Brakel, Kushwaha, Snodgrass, and myself that at least begins to point in the right direction for finding such a mechanism. In a study of spider phobics who were given a signal detection test for drawings of spiders below the objective detection threshold we found that, unlike snake phobics who were shown the same spider stimuli, the brain response at the *vertex* was *greater* than at the *occiput*. Secondly, the same HOQ correlated *negatively* with spider detectability for the spider phobics but not for the snake phobics. Interestingly, the same negative correlation was found for self-rated anxiety. Putting the findings together we can surmise that the repressive activity begins downstream from its perceptual registration. Not too surprising, but a start. Perhaps the newer imaging procedures such as the fMRI would make it possible to identify with more precision the brain elements involved. We are at present exploring the possibility of a related study using the pop/look paradigm to investigate unconscious inhibition in collaboration with Daniel Glaser at the Institute for Cognitive Neuroscience, University College, London.

3. QUALITATIVE DIFFERENCES BETWEEN CONSCIOUS AND UNCONSCIOUS PROCESSES

In the subliminal research literature the interest in qualitative differences between conscious and unconscious processes emerged from a methodological dilemma: since most studies as already noted were conducted at the subjective threshold so that phenomenal consciousness could not be ruled out, the argument is made that if qualitative differences were found between the presumed subliminal effect and conscious performance then it could be inferred that the effects were truly subliminal. As Snodgrass (2005a) has pointed out, the argument is far from convincing because the qualitative differences could characterize phenomenal as opposed to reflective awareness. And in fact, in a study in which the role of reflective awareness wss heightened qualitative differences disappeared (Visser & Merikle 1999). The qualitative differences I will now describe are primarily of theoretical relevance, although they also possess methodological implications.

In order to better appreciate the nature of these qualitative differences I will need to sketch briefly a theoretical frame of reference based largely on the thinking of Charles Fisher, a pioneer in psychoanalytically inspired subliminal research. In an article published in 1957, Fisher presented a model in which there was one essential assumption: That all stimuli registered initially at a preconscious level. Certainly since then a considerable amount of evidence has accumulated in support of this assumption Cognitive processes such as categorization, for example, can occur prior to consciousness. Fisher contended that once registered preconsciously that a stimulus can undergo different fates: 1. most usually it is delivered directly into consciousness with very little modification, in particular if it is of sufficient intensity, 2. it could remain pre-conscious if, for example, it were not of immediate attentional interest, 3. it could become associated with anxiety-arousing significance because of some link to unconscious conflict and be inhibited from entry into consciousness. These three factors, stimulus intensity, attention, and unconscious significance act independently of each other and can strengthen, weaken, or cancel each other. Thus, as Brakel (1989) has illustrated from clinical experience and as Spiegel, Cutcomb, Chuen & Pribram (1985) have shown through hypnosis, a highly intense stimulus which would ordinarily quickly become conscious can be kept from consciousness by a more powerful unconscious motivation and result in a negative hallucination- the absence of conscious perception in the face of a clear sensory presence.

When a subliminal stimulus registers preconsciously it lacks sufficient stimulus intensity to follow immediately the path to consciousness. Yet it is has entered through a perceptual pathway, is encoded, and much evidence indicates can have various indirect effects on consciousness. Perhaps the clearest examples are provided by priming effects. In some instances, as in our pop/look experiments, stimulus itself enters consciousness in the form of a correct guess. It is this latter finding which is of special interest with respect to qualitative differences. Although it is true that the correct guess enters consciousness, there is one significant quality it lacks – *that of being experienced as a perception even though it has registered through a perceptual channel.* Yet only its contents enter consciousness and in different form..

One might ask why the subject does not have a perceptual experience, exclaiming, "Oh, yes that was the word just flashed". The answer appears to be obvious, but is nevertheless of central importance: There is no such conscious experience of a perception because it was never in consciousness to start with. From this it is possible to infer that the process of becoming conscious contributes to an important and necessary function generally overlooked in the literature on consciousness. In order to appreciate the nature of this function I will need to distinguish between a *representation* and what I will call its *mental vehicle*. A representation of a cat can appear in consciousness as a perception, a memory, an image, a thought, part of a fantasy, and so on. These are all mental vehicles in which, if you like, the same representation can take a ride at different times. Remarkably, we do not seem to have any trouble most of the time in distinguishing a representation from its vehicle. We know when we are experiencing a perception, an image, a memory, or a thought even though the content may be identical. Put differently, in each case the vehicle as a mental act is *about* something in Brentano's sense of *intentionality*, that something is the representation or content. What I am proposing is that the process of becoming conscious confers this vehicular tag. When stimuli enter preconsciously and therefore have not as yet participated in the process of becoming conscious they lack that tag and thus can readily be confounded with images, memories, thoughts, or other mental vehicles. The experience of the correct guess in subliminal experiments comes as something equivalent to a thought, or a possibility, not of a perception or memory. Given the task it is not difficult to see that this thought is called a guess.

There is one other related hypothesis: Even though once conscious a representation has been tagged with its mental vehicle it can lose its tag if it drops out of consciousness as a result of repression. In the repressed, or dynamic unconscious, representations have lost their vehicle tag and thus for example, perceptions and memories can be confounded accounting for the apparent timelessness of the dynamic unconscious. This theory has been presented at greater length elsewhere (Shevrin 1986, 1998).

With these conceptual tools in hand I would like now to address the issue of qualitative differences between conscious and unconscious processes as they apply to the research I have thus far cited.

I have already described findings bearing on one such potential qualitative difference: there is a different relationship between conscious and unconscious processes as a function of stimulus detectability. Here we come across the first stumbling block. I have argued that whether d' is above or below the mean in experiments below the objective detection threshold, the stimuli are still unconscious, or perhaps more consistent with the terminology just introduced, are still preconscious. Then why should above and below d' scores relate differently to conscious guessing? And moreover, why should they be correlated with repression? Presumably according to the hypotheses just advanced they all lack the conscious memory tag.

The problem can be addressed by conceptualizing conscious tagging as a process that is continuous rather than categorical. As d' increases the representation involved becomes closer to the threshold of consciousness, and takes on more of a perceptual tag. Prior to consciousness the tagging is not either or, but one of degree. Once conscious however it becomes fully tagged. An analogy from the old days of home photography might be helpful. In developing a print the negative was first bathed in a solution that begins to bring out the light and dark exposed areas so that one can begin to make out forms, but if one were to take the negative out of the solution at that point the emerging light and dark patterns would gradually disappear. Instead one bathes the developed negative in a fixing solution that completes the printing process. Similarly as d' progresses above zero it is developing as a perception, but only after it enters consciousness is it fixed or tagged as a perception. It is also possible to imagine that the binding process by which individual features are bound together to form an object may be another analogue if not a case in point for the development of a perceptual tag.

For the lookers in the look strategy to the extent that d' is above the mean to that extent the correct stimulus is assuming more of a perceptual character. Since at no point has thelooker had a fully conscious experience of the stimuli, this is at variance with the looker's conscious experience that there was nothing to be see. The correct response is rejected. On the other hand, to the extent that d' is below the mean it lacks a perceptual tag and presents no inconsistency with what is experienced consciously. It is therefore more likely to be accepted as a guess by lookers.

For the repressors, who responded in the same way as lookers, the outcome is the same but the means used are different. This difference can best be illustrated by the result from the social phobia experiment. In that experiment repressors showed the greatest difference in favor of subliminal over supraliminal differentiation of unconscious conflict words. When the unconscious conflict words were presented supraliminally they possessed sufficient stimulus intensity to enter consciousness as perceptions. But the brain responses tell us that they were not differentiated from the other three sets of words as they had been subliminally. They no longer were experienced as different from the other words.

As suggested previously repression removes the mental tags from representations. Once this happens a different mode of thinking becomes possible. Since perceptions, memories, fantasies no longer exist as discrete mental categories it is possible for representations initially linked to a particular mental category to meld and flow together creating the conditions for condensations and displacements, both mechanisms of what Freud referred to as the primary process. In another series of experiments we have identified two formal properties of categorization which distinguish between this primary process melding of representations and the more discrete, rational form of categorization characterizing what Freud referred to as the secondary process. It would take me well beyond my time allotment to describe these studies and their implications for understanding further qualitative differences characterizing the interaction of conscious and unconscious processes. Fortunately, my colleague Dr. Brakel has presented this research in some detail in her chapter for the volume *Psychanalyse im Dialog der Wissenschaften. Anglo-amerikanische Perspektiven* (Brakel 2004, in Giampieri-Deutsch 2004).

4. CONCLUSIONS

I will now return to the four propositions stated at the beginning of this chapter and see how the findings described in the body of my contribution support them.

1. Consciousness cannot be fully understood without knowledge of unconscious processes.

The findings most relevant to this proposition are provided by the social phobia study in which the patients during the supraliminal presentation of the unconscious conflict words consciously experience simply perceiving words familiar to them. But their brain responses and the HOQ tell us a different story. The unconscious conflict words are being less well differentiated from the other words than they had been subliminally, and the extent to which this happens is a function of an unconscious inhibition. The state of affairs is often the case clinically, in particular during the early phases of treatment when neither patient nor analyst is aware of the special significance of certain communications, although the same inhibitory processes are at work as revealed later in the treatment.

2. Unconscious processes are complex, representational, highly interactive with conscious processes and not dispositional or latent.

Again the social phobia study can best speak to this proposition. An unconscious conflict is a complex organization of perceptions, memories, feeling, fantasies and desires centered around significant people both past and present in the individual's life. In these respects they are as complex and representational as conscious processes. The unconscious conflict words selected by the psychoanalysts constitute a *sampling* of these complex representations and their interrelationships. They are thus representations in their own right as well as complex. Because of the persistence of these unconscious conflicts hypothesized to cause the social phobic symptom, they are active and influence ongoing conscious processes. This is evidenced by what has already been said with respect to the first proposition. When the complex representations of the unconscious conflict are presented supraliminally, their unconscious significance influences the fate of these representations in consciousness.

3. The interaction between conscious and unconscious processes can best be understood when person and context are taken into account.

Evidence for this proposition is provided by both the pop/look studies and the social phobia study. If context in the form of different strategies and individual differences in the form of strategy preferences had not been included in the series of pop/look studies no subliminal effects would have been found. In the social phobia study this is further supported by the use of unique, personally meaningful words, a considerable departure from most cognitive experiments in which generally the same stimuli are used for all subjects. Nevertheless, only through the use of these personally meaningful words was it possible to discover a functional relationship true for all of the subjects. The time to consciousness study also revealed the importance of individual differences; once taken into account an interesting finding emerged concerning the relationship to repression.

4. Quantitative and qualitative differences between conscious and unconscious Processes.

The results bearing on the relationship between stimulus intensity and different thresholds indicate that quantitative and qualitative factors are closely intertwined. At the objective detection threshold whether one is above or below the d' mean, an unconscious quantitative factor, determines whether one will facilitate or inhibit under certain conscious task instructions, a qualitative factor. It is also important to keep in mind that while measured as a quantity, d' itself is associated with an important qualitative feature. As d' rises above zero it assumes more of a perceptual tag; as d' falls below zero it ends to lose its perceptual tag. At higher and higher levels of d' the stimulus begins to acquire a perceptual character until at the subjective threshold they enter phenomenal consciousness as perceptions. It was further suggested that consciousness confers a mental vehicle tag on a representation so that the representation is identified as a perception, memory, image, fantasy, etc., and that repression undoes this tagging so that repressed representations can flow and meld across mental vehicles producing the condensations and displacements identified by Freud with the primary process.

These propositions taken together make possible a different approach to conceptualizing the nature of consciousness and its relationship to unconscious processes from that currently prevailing in much research and theorizing in the field. For example, from this standpoint, unconscious effects are neither simple nor evanescent as has been advanced recently. As the audience seated in front of the visible theater we are challenged to figure out what is happening on the other stage if we are to understand what is going on before us. As a psychoanalyst this has always been the clinical challenge. Fortunately we now have experimental methods taking us beyond the limitations of the clinical setting that allow us to look more directly on what is happening on the invisible stage and in this chapter I have tried to emphasize what we have learned from these more direct observations.

REFERENCES

Ayckbourn, A. (2000). *Home & Garden*. London: Faber & Faber.

Baars, B (1997). *In the Theater of Consciousness: The Workspace of the Mind*. New York: Oxford University Press.

Bernat, E., Bunce, S. & Shevrin, H. (2001). Event-related potentials differentiate positive and negative mood adjectives during both supraliminal and subliminal visual processing. *International Journal of Psychophysiology* 42: 11–34.

Bernat, E., Shevrin, H. & Snodgrass, M. (2001). Subliminal visual oddball stimuli evoke a P300 component. *Journal of Clinical Neurophysiology* 112: 159–171.

Block, N. (1995). On a confusion about a function of consciousness. *Behavioral and Brain Sciences* 18: 227–287.

Bock, G.R. & Marsh, J., ed. (1993). *Experimental and Theoretical Studies of Consciousness*. CIBA Foundation Symposium 174. Chichester: Wiley.

Brakel, L.W. (1989). Negative hallucinations, other irretrievable experiences and two functions of consciousness. *International Journal of Psychoanalysis* 46: 461–479.

Brakel, L. (2004). Der Primärvorgang: Experimentalle Belege für A-rationale Organisationsprinzipien und einen philosophischen Fall von A-rationaler Referenz. In *Psychoanalyse im Dialog der Wissenschaften. Vol. 2. Angloamerikanische Perspektiven*, ed. P. Giampieri-Deutsch. Stuttgart: Kohlhammer, pp. 143–167.

Brewer, M.B. & Luce, R.D. (1998). Basic research in cognitive science. *American Psychological Society Observer*, Report 6: 3–39.

Caine, T.M. & Hawkins, L.G. (1963). Questionnaire measure of the hysteroid/obsessoid component of personality: The HOQ. *Journal of Consulting Psychology* 27: 206–209.

Edelman, G. (1987). *Neural Darwinism: The Theory of Neuronal Group Selection*. New York: Basic Books.

Fisher, C. (1957). A study of the preliminary stages of the construction of dreams and images. *Journal of the American Psychoanalytic Association* 5: 60–67.

Friston, K. J. (2000). Imaging neuroscience: 1. Neuronal transients and nonlinear coupling. *Philosophical Transactions of the Royal Society of London B* 355: 215–236.

James, W. (1890). *Principles of Psychology*. New York: Holt

Kandel, E. R. (1999). Biology and the future of psychoanalysis: A new intellectual framework for psychiatry revisited. *American Journal of Psychiatry* V.156(4): 505–524.

Libet, B., Alberts, W.W., Wright, E.W., Delattre, L.G., Levin, G. & Feinstein, B. (1964). Production of threshold levels of conscious sensation by electrical stimulation of human somatosensory cortex. *Journal of Neurophysiology* 27: 546–578.

Ludolph, P.S. (1981). The dissociative tendency, its relationship to personality style and psychopathology. Unpublished doctoral dissertation. Ann Arbor: University of Michigan.

Shevrin, H. (1986). A proposed function of consciousness relevant to theory and practice. Presented at the American Psychological Association Convention, Washington, D.C., August 22, 1986.

Shevrin, H. (1998). Why do we need to be conscious? A psychoanalytic answer. In *Advanced Personality*, ed. D.F. Barone, M. Hersen & V.B. Van Hasselt. New York: Plenum Press.

Shevrin, H., Bond, J.A., Brakel, L.W., Hertel, R.K. & Williams, W.J. (1996). *Conscious and Unconscious Processes: Psychodynamic, Cognitive, and Neurophysiological Convergence.* New York: The Guilford Press.

Shevrin, H., Ghannam, J. & Libet, B. (2002). A neural correlate of consciousness related to repression. *Consciousness and Cognition* 11: 334–341.

Shevrin, H., Williams, W. J., Marshall, R.E., Hertel, R.K., Bond, J.A. & Brakel, L.W. (1992). Event-related potential indicators of the dynamic unconscious. *Consciousness and Cognition* 1: 340–366.

Snodgrass, M. (2005a). Disambiguating conscious and unconscious influences: Do exclusion paradigms demonstrate unconscious perception? *American Journal of Psychology*, in press.

Snodgrass, M. (2005b). *Unconscious Perception: Theory, Method, and Evidence.* Philadelphia: John Benjamins, in press.

Snodgrass, M., Bernat, E. & Shevrin, H. (2005). Unconscious perception models. *Perception and Psychophysics*, in press.

Snodgrass, M., Shevrin, H. & Kopka, M. (1993). The mediation of intentional judgments by unconscious perceptions: The influence of task strategy, task preference, word meaning, and motivation. *Consciousness and Cognition* 2: 169–193.

Spiegel, D., Cutcomb, S., Chuen, R. & Pribram, K. (1985). Hypnotic hallucination alters evoked potentials. *Journal of Abnormal Psychology* 94: 249–255.

Van Selst, M. & Merikle, P. (1993). Perception below the objective threshold? *Consciousness and Cognition* 2: 194–203.

Visser, T. & Merikle, P. (1999). Conscious and unconscious processes: The effects of motivation. *Consciousness and Cognition* 8: 94–113.

III. DRIVE THEORY AND PRIMARY PROCESS: A PHILOSOPHICAL ACCOUNT

LINDA W. BRAKEL

1. INTRODUCTION

1. 1. DRIVES DEFINED AND DESCRIBED

Drive theory in psychoanalysis concerns the "instinctual drives". Currently there is much controversy regarding the relation between instinctual drives and affects. Although for some theorists these are placed in a single category, as will be plain, the very philosophical analysis I am undertaking presupposes what amounts to a necessary and fundamental difference between drives and affects. Much of the account of drives provided in this article concerns the nature of drive objects; while clearly the very notion of an "affect object" is incoherent. Thus to my mind drives and affects must belong to two distinct categories. While this confounding of conceptual categories might be the latest problem drive theory is facing, it has not been its only one. At various times and in various quarters there have been great rhetorical debates and confusions about Freud's very use of the term. For the purpose of this article however (because the terminological issue is not relevant to the philosophical analysis) I will use what has become the standard reading (Nagera 1969, 1970) starting with Freud's (1905d) comments from *Three Essays on Sexuality*:

"By an ,instinct' [instinctual drive] is provisionally to be understood the psychical representative of an endosomatic [...] source of stimulation [...] the concept of instinct [instinctual drive] is thus one of those lying on the frontier between the mental and the physical. The simplest and likeliest assumption [...] would seem to be that in itself an instinct is without quality, and, so far as the mental life is concerned, only to be regarded as a measure of the demand made upon the mind for work. What distinguishes the instincts [instinctual drives] from one another and endows them with specific qualities is their relation to their somatic sources and to their aims " (Freud 1905d, p. 168).

After using almost identical words to define instinctual drives, Freud (1915c) in *Instincts and Their Vicissitudes* elaborated four terms needed to understand the concept. In addition to the "source" and "aims" of an instinctual drive, described in the earlier work, he added the "pressure" of a drive and its "object" (p. 122–3):

"By the pressure of an instinct we understand its motor factor, the amount of force or the measure of the demand for work which it represents [...] The aim of an instinct is in every instance satisfaction [...] obtained by removing the state of stimulation at the source of the instinct. But although the ultimate aim of each instinct remains unchangeable, there may yet be different paths leading to the same ultimate aim; so that an instinct may be found to have various nearer or intermediate aims [...] The object of an instinct is the thing [...] through which the instinct is able to achieve its aim. It is what is most variable about an instinct and is not originally connected with it, but becomes assigned to it only in consequence of being peculiarly fitted to make satisfaction possible. The object is not necessarily extraneous; it may equally well be a part of the subject's own body. It may be changed any number of times in the course of the vicissitudes which the instinct undergoes [...] It may happen that the same object serves for the satisfaction of several instincts simultaneously" (Freud 1915c, p. 122–3).

By the source of an instinct is meant the somatic process which occurs in an organ or part of the body and whose stimulus is represented in mental life by an instinct.

In sum then, a drive is the mental representative of a physiologic need state such that a certain pressure or quantitative demand for work, having a certain quality varying with particular drives, is placed on the psychological apparatus toward fulfillment and thereby cessation of the particular drive. Each drive in addition to its pressure, has a source (in the body), an aim (the distal part of which is always fulfillment, the proximal part varies with each drive and is the mode to achieve fulfillment), and an object. The object is that thing or those things or parts of things (which can include self or other or part thereof) through which satisfaction can be obtained. The source, aim, and object of a drive determine its qualitative nature. The pressure of a drive, mostly originating from its somatic source, constitutes the quantitative contribution.

1. 2. An Example: X's Oral Drives

X is a baby, and X is hungry. X has a number of oral drives – to suck, swallow, and to feel the pleasure associated with pleasant taste and temperature and the comfort of gradual gastric fullness replacing emptiness. The somatic sources are falling blood sugar, hypothalamic regulation and signaling, gastric stirrings, and sensations in the mouth. The aims are, distally to have this drive state reduced, and proximally to get fed. The object is whatever will satisfy these oral drives. Food not separable from whomever (or whatever or part of whomever or whatever) provides it, is an obvious object; but so is the pacifier, the rattle, and part of the blanket – whatever is good to suck on. Thus for Baby X the object of these oral drives is X's mother's left breast and nipple and her right breast and nipple and a bottle and nipple, and the milk, and her face, and how it feels being held and fed, and how her face and breast (or bottle) look, and the pacifier and the hand of its provider, and the blanket edge, especially how it feels in X's mouth – any part of this and all of it together.

2. A PHILOSOPHICAL ANALYSIS OF DRIVES

2. 1. DRIVES AND PROPOSITIONAL-ATTITUDE PSYCHOLOGY

Drives do not fit well into the propositional-attitude psychology favored by philosophers to account for motivation and intention. Desires, in contrast, as the conative propositional attitude type often paired with beliefs (a cognitive propositional attitude type), fit readily. What is the difference? First here are some important similarities. Both drives and desires can be used in conjunction with behavior to infer the content of various cognitive propositional attitudes. For example, if Mr. A says he desires a cool drink, then is seen walking toward a water-fountain, and then bends to drink, one can infer that he believes that water-fountains are sources of cool drinks and that he supposes (another cognitive propositional attitude) that this particular water-fountain is operational. Under some conditions the very same cognitive propositional attitude contents can be inferred from beliefs about the state of Mr. A's drives along with certain actions he performs, without his expressing a desire. Suppose, for example it is learned that Mr. A. has been deprived of fluids for some time. This state will, assuming Mr. A. is normal, lead to heightening of various oral drives and the very same inferences regarding the propositional attitudes associated with his ensuing acts follow. Conversely drives and desires can themselves be inferred given various cognitive propositional attitudes. Mr. A. starting to drink from a water-fountain usually implies that he believes that water-fountains are sources for cold water, and that he supposes that the particular water-fountain he is about to drink from works. From here it can be inferred that he is drinking because he desires a cool drink, likely in part owing to various oral drives that are active. Now this sort of belief/desire propositional attitude analysis can go wrong. Mr. A. can be a water-fountain inspector, with oral drives that are well satisfied, no desire whatever for cool drinks, and a belief that the water-fountain in question is in need of repair. But most of the time the sort of analysis performed on Mr. A. will go through, even dealing with drives. So what is the difference between drives and desires? And will we find that this difference accounts for why drives don't fit the standard propositional-attitude psychology analysis? The answer to the second question is yes.

Here is what is different. Whereas with any desire (or belief) its object can be picked out by a mental representation the content of which is determinate, no such singularly determinate content will do for the object of a drive. The representation of the objects of drives look radically indeterminate in content (see Velleman 2002, p. 90). Mr. A's desire for a cool drink has as its object "cool drink" and he has a mental representation of this. Whatever the nature of mental representations, Mr. A's mental representation of cool drink, refers singularly to cool drinks. Turning now to the object of Mr. A's oral drives, those likely driving his desire to seek a cool drink, there is no such straightforward analysis. The object of these drives, hence the content of the mental representation picking out the object, cannot be specified as singular and determinant, anymore in Mr. A's case than for Baby X above. The object can be approximated as some sort of amalgam of objects and parts of objects, all and whatever of which will bring forth a satisfied feeling in Mr. A's mouth. But any mental representation of this sort of drive object will seem "either radically indeterminate in content or as playing a non-standard role" (Velleman 2002, p. 90). Velleman (2002, p. 91) describes such representations as "far too vague to be expressed in the concepts with which we consciously reason." He, in discussing the object of aggressive drives, continues (p. 91):

"If [...] we insist on framing a written or spoken 'that' clause to express the content of aggression, we shall have to concede that what the [aggressive drive] attitude can [...] be satisfied by [...] [is] not only the literal truth of the clause but also indefinitely many other outcomes related only by analogy, by metaphorical similarity, or by other mental associations of an open-ended variety."

Velleman concludes (p. 91): "Either way, propositional-attitude psychology will not afford the same computational advantages in this case as it does in the case of ordinary beliefs and desires, whose tendencies [...] to be satisfied can be summed up in sentences of ordinary language."

2. 2. Are drive aims singular and determinate?

That drives have objects that cannot be represented in a univocal, determinant way is central to drives not admitting of a standard propositional-attitude psychology analysis. (More will be said about the non-singular nature of drive objects in the next subsection.) But is the same absence of singular determinancy true of the aims of drives? In Freud's (1915c, p. 122) basic definitions, it is clear that the aim of any drive will consist of two parts: the "ultimate and unchangeable" aim of ceasing the state of stimulation through discharge and satisfaction, and the "different paths" such that a drive may have "various nearer or intermediate aims". Given this compound aim for any drive, is there determinancy of the nearer and intermediate aims of a drive? Anna Freud (1936, p. 175) describes the aims of drives as easily modified, transformed, and displaced one for another. In a later work she states: "An individual who finds the path to an instinctual [drive] aim of adulthood blocked by internal or external obstacles can respond to such a frustration by reverting to childhood forms of satisfaction" (1977, p. 248). And she notes this can go in the opposite direction too: "under the influence of the superego, primitive [drive] impulses are transformed, inhibited in their aims, and the drive energies belonging to them deflected to moral, ethical, or socially higher aims." (p. 247). All of Freud's cases (See especially Dora in Freud 1905e; Little Hans in Freud 1909b; The Rat Man in Freud 1909d; and The Wolfman in Freud 1918b) show evidence of displaced aims. Further, his understanding of regression, symptom formation, and sublimation, despite other changes in his conceptions of these, all remained dependent on what for him was obvious: the aims of drives could be readily displaced (1900a, 1915c, 1915e, 1926d, 1940a).

Although it may seem that the foregoing suggests indeterminacy, I contend that in fact, the very transformability or displace-ability of the aims of drives, suggests that drives have original aims from which there can be displacement. There is support for this understanding from Freud (1905d, p. 168) in his *Three Essays on Sexuality*. In a passage stating that all instinctual drives have the distal aim of discharge and satisfaction, but differing proximal aims, he states that: " What distinguishes the instincts one from another and endows them with specific qualities is their relation to their somatic sources and to their aims." Discussing Freud's allusion to the "nearer and intermediate aims" Rapaport (1967, p. 831) comments: "Freud used the term 'aim' in two different senses. One of these is the tendency toward discharge; the other [...] is not the universal discharge tendency itself, but something more specific." The more specific quality of aims, is determined by the "instinctual mode" (according to Rapaport, p. 832, this is Erikson's concept; Erikson 1950), each mode associated with the source of an instinctual drive. Thus as a child gaining anal sphincter control is in the anal phase, the anal drives are associated with aims whose mode concern retaining and expelling. Anal phase drives of course far outlast the anal phase of development, and drives with aims to retain (money, for example) or expel (anger, troublesome others) are commonplace; but no moreso than displacements from these anal phase aims. For example the aim to retain (money) can be displaced onto the perhaps more acceptable aim of constructing a bank. Thus it is in the more specific sense of aim – the proximal aims with their particular instinctual modes – that there are original aims and then displacements from these originals. As Rapaport put it (p. 831): "Clearly, in this second sense of the term, *the aim* [my emphasis] can be restricted, displaced and substituted for."

2. 3. Drive objects are not singular. Must they be indeterminate?

The threat of indeterminancy for drive objects cannot be resolved so easily. While the case has been made for drives having original singular aims from which displacements occur, there is not a parallel case for drive objects. Returning to Freud's (1915c, pp. 122–3) definition of instinctual drive object, he states: "It is what is most variable about an instinct and is not originally connected with it, but becomes assigned to it only in consequence of being peculiarly fitted to make satisfaction possible." With drives possessing no specified original objects through which to gain discharge and satisfaction, in the role of drive object any person or material object or part thereof which fits a drive in order to make satisfaction possible is as good as any other person, material object, or part thereof.

Velleman (p. 90), in discussing his own aggressive drive gives a convincing description of the non-singularity of drive objects. While performing various kicks in Tae Kwon Do exercises, he found that his "…aggression could be turned on virtually any solid object, including any person who happened to be my assigned opponent."

Rapaport (1967) also recognized the non-singularity of drive objects, claiming that the assignment of object to drive in consequence of good fit for gratification implied "a coordination of the instinctual drive discharge with a definite (even if broad) range of objects, and an equifinality in regard to this range of objects" (p. 878). (Rapaport's assertion that the range is "definite" will not suffice to ensure determinancy. It does however suggest that he wanted this assurance too.) Rapaport (p. 878) further holds that that what constituted the range of well-fitted objects was likely "phylogenetically determined". Earlier (p. 833) he had noted a parallel between Freud's set of drive objects and the "releasers" in Lorenz's system of innate releasing mechanisms (the most well known of which is imprinting). Rapaport reports that Lorenz (1950) and Timbergen (1951) found that the releasers were not singular objects. Rather that they were "aggregates of stimuli, only some of which are indispensable for effecting adequate release, and various combinations of which can effect adequate release" (The quotation is from Rapaport 1967, p. 833).

This finding accords well with the above description of the object of Baby X's oral drive. This object was said to be any amalgam or part of the provider of satisfaction and/or the material object satisfying Baby X's oral drive: X's mother's left breast and nipple and her right breast and nipple and a bottle and nipple, and the milk, and her face, and how it feels being held and fed, and how her face and breast (or bottle) look, and the pacifier and the hand of its provider, and the blanket edge, especially how it feels in X's mouth – any part of this and all of it together.

Two things can be seen in all of these examples of drive objects: 1. There is no primacy or originality of one part of these objects over another. In other words, displacement from some original is not taking place here for drive objects as it is for drive aims; 2. These objects do not appear as singular and determinate, at least not in any ordinary way. Instead they are either amalgams of parts of persons or items (as in Baby X and in the animals studied by Lorenz and Timbergen); or a set of items linked as a set by some contingent feature as in Velleman's example of "any solid object" which happened to be available.

These mysterious objects can be understood however, once they are seen as being composed of elements comprising an a-rational primary process category. Primary process categories are collections grouped together, not at all randomly, but lawfully; not according to rational principles, but on the basis of the associative, a-rational principles of the primary process (see Freud 1900a, 1915e, 1926d, 1940a). Drive objects are constituted as such categories whose elements cohere owing to different primary process principles. For example, there is one type of primary process category held together due to a shared feature or attribute, one that need not be an essential feature. And indeed, there are drive objects which do seem to be constituted in just this way – take for example, Lorenz's ducklings who will imprint onto any moving item of a certain physical size, Velleman who would kick any solid object, and Baby X who sucks on anything that is available to suck. Another primary process category links things that are spatially contiguous. And we can see this sort of drive object too – i.e., Baby X's drive object consists of mother's left nipple and the milk, father's hand and the nipple on the bottle and the milk, the pacifier and the face of a parent, etc. (The topic of primary process categorizations has been addressed much more fully in Brakel 2002, 2003, 2004).

2. 4. Can primary process objects be represented? How?

Drive objects can be better understood when they are seen as objects that cohere on the basis of primary process principles. But there is still the problem of indeterminancy. How can such objects be represented? Given that the standard propositional-attitude psychology analysis used to individuate contentful representations of the objects of propositions will not work – and it will not work because drives are not standard propositional attitudes with singular, representable objects – is there some analysis that can fix the content of these objects?

Certainly fixing content with the normative assumption of interpretative holistic rationality (as do the attributionists, one group using propositional-attitude psychology), or applying any sort of rationality based fixing will not work. (For attributionist and other rationality based views, see especially Davidson 1980, 1984; Cherniak 1981; Stich 1983; and Dennett 1978, 1987.) How could this work when we are dealing with a-rational, albeit lawfully organized categories? What would work is an account with a different normative assumption than that of interpretable holistic rationality and yet an account not susceptible to the indeterminancy problem. These requirements can be met using the proper-function naturalism developed by Ruth Millikan (Millikan 1984,

1993) in which the normative assumption of interpretative success and rational explanation is replaced with the normative assumption of selective success and evolutionary explanation.

I will briefly describe this fitness-based normativity. Then in keeping with the terms of this analysis, I will claim that indeed primary process based drive objects do have fixed, albeit non-singular contents insofar as they have a proper function, one which under certain conditions leads to fitness success. (What follows is an application of an argument concerning the proper function and fitness success of a-rational primary process phantasies, appearing in Brakel 2002).

2. 5. Fitness based normativity: A proper function account of representations of ordianry objects

In order to demonstrate selective fitness normativity, let's start with something far from our area of concern – sweat glands. The function of sweat glands is to secrete sweat. But their "proper-function" entails what Millikan (1993) terms "Normal" conditions. Normal conditions are those in which properly functioning sweat glands contributed to the survival of animals possessing them and thereby increased the selective success of those animals. Thus sweat glands functioning flawlessly to produce sweat when there is a tumor at the temperature-sensing center are not performing their "proper-function" because they are not functioning under "Normal" conditions – i.e., those conditions contributing to the animal's selective reproductive advantage. Only those sweat glands producing sweat when the body is overheated are sweat glands properly-functioning under Normal conditions.

Now let's move closer to our area of interest, beliefs and their objects. Not all beliefs are properly-functioning under Normal conditions. Normal conditions, those which conferred selective fitness, are those in which the truth conditions of the representation (the object of belief) match those obtaining in the world. In other words only true beliefs are those properly-functioning under Normal conditions (Millikan 1993, p. 73). But note that in Millikan's program there is nothing a priori or necessary in the object of beliefs matching truth conditions in the world or in believer coming by the object of their beliefs rationally. Rather true beliefs and the rationality they usually entail are contingently determined biological norms. Yes, it happens to have been the case that our ancestors capable of having rationally mediated true beliefs, and exercising this capacity just frequently enough, had selective advantages such that the mechanisms supporting true-belief generation were reproduced. But that the Normal conditions for beliefs are those in which the truth conditions for the representations are constituted by the truth conditions of the world, is entirely a contingent matter. Note also that the Normal conditions need not be the typical conditions. Provided that there are enough Normal properly-functioning beliefs to fix content via fitness success and selective normativity, instances of abNormally functioning beliefs, i.e., false beliefs, incomplete beliefs, etc. will also have content.

2. 6. Fitness based normativity: A proper function account of an unusual type of representation

Closer still to our interest in primary process drive objects, let's turn to toads. Toads can be used to demonstrate 1. representational content without any rationality; 2. abNormal instances that are nonetheless contentful; 3. representations which have fixed determinate content despite initial appearances to the contrary. Toads, Millikan (1993, p. 109) claims, have contentful representations of bugs, despite their incapacity to distinguish bugs from lead pellets. Toads have a powerful set of reflexes such that when bugs or lead pellets are within striking distance tongue protrusion, item capture, and item swallowing ensue in rapid sequence. Lots of bugs and lots of pellets are swallowed indiscriminately; and yet Millikan claims that toads have a determinate representation of bugs. Here is how. Although a toad cannot differentiate a bug from a lead pellet, presumably producing indistinguishable visual representations, only those representations which have represented actual bugs are the result of properly-functioning Normal condition representations. Why? Because only actual bugs have contributed to toad nutrition and thereby to toad reproductive success. And what about those representations that represent actual lead pellets? They are not properly-functioning Normal condition representations of bugs; nor are they any sort of representations of lead pellets; they are abNormal representation of bugs! The content "bug" is fixed in a determinate fashion according to selective fitness normativity.

2. 7. Fitness based normativity: A proper function account of primary process drive objects

A proper function account is given in two steps. First the proper function is set out; then the Normal or fitness enhancing conditions are proposed. So let's begin with the proper function for the object of drives.

2. 7. 1. Proper function

Obviously, any proper function of the drive objects must ensure that the drive aims – both the aims specific to particular drives, and the ultimate aim of drive discharge – are satisfied, with the result that the demand for work the somatic source places on the mental apparatus is considerably diminished. Less obvious, but no less important, the proper function of drive objects constituted by primary process organized sets is two-fold. First the proper function is that such objects are constituted by elements connected in a primary process fashion. The resultant set is not in any way random, but coheres according to primary process principles – association by contiguity in time or space, and association by the similarity of inessential features. Second the proper function of such primary process drive objects allows that this sort of diversity of drive satisfaction is tolerated. Rapaport (1967, p. 833) suggests as much: "The coordination of instinctual drive and object is assumed to be innate, i.e., given by evolution. While this coordination is rigid in lower-order animals, it is relatively flexible in man." (The adaptational relations between drive, drive object, and environment is taken up in detail in Hartmann 1958). Rapaport continues by noting that there can be considerable "variability in the object of the instinctual drive". And finally Rapaport even challenges the rigidity of drive and drive object in animals. Citing the findings of Lorenz and Timbergen (quoted above) in which the so called "innate releasing mechanisms releasers" (drive object equivalents) are not singular and complete objects, but rather "aggregates of stimuli, only some of which are indispensable for effecting release, and various combinations of which can effect adequate release", Rapaport concludes: "The concept of releaser, like Freud's instinctual-drive object, implies some flexibility of the relation between the innate releasing mechanism and the releaser."

2. 7. 2. Normal conditions

Having specified the proper function for drive objects constituted by primary process sets, the next step is to find those conditions under which the proper functioning of these drive objects could lead to fitness success. If found, these will be the Normal conditions, that allow the primary process drive objects to have fixed determinant content, without being univocal and singular. This is a surprisingly easy task. Let's look at the consequences if primary process amalgams or sets did not comprise drive objects. Suppose Lorenz's ducklings could imprint uniquely only onto their mothers. Yes, if you are a duckling it is ideal to imprint onto your mother; but suppose she is gone or sick? Isn't better to be able to imprint onto some other conspecifics (or even onto a researcher) and have a chance thereby to survive and reproduce? There seems to be clear selective fitness advantage in the diversity of the type of primary process organized "releasers" ducklings possess. Similarly with human babies, if Baby X's oral drives could only be quelled with mother's milk from her nipples, Baby X would likely be hungry more of the time, certainly discontent much of the time, and plausibly less likely to grow up healthy enough to reproduce. Jumping ahead developmentally, suppose drives from the Oedipal phase could have but a unique, singular type object. Wouldn't Oedipal phase resolution be even harder than it is? Wouldn't choices for mates be severely curtailed, at the detriment to both individual and species? If we turn to aggression, the same potential benefit for primary process objects can be seen. Suppose for some people the "aggressive drive" had as its object only conspecifics, those vying for a mate, those attacking first, etc. These people would be at a selective disadvantage with respect to nourishment to those whose aggressive drive also included prey and predator animals. The fitness disadvantage would be no less for people whose aggressive drive had the singular object of big predators. For these people would have trouble defending themselves and progeny against attacks from other people, as well as from noxious, but small animals (like snakes and spiders). They might also be at a calorie disadvantage, finding less aggressive drive pleasure in fishing and small game hunting.

In the last two subsections I have made a case for a proper function account of drive objects constituted by primary process amalgams or sets. There are criteria for the proper function of primary process set (or amalgam) based drive objects – namely, that the set of elements comprising the drive object is organized according to primary process principles; and that the resultant drive object allows diverse gratification. And there are Normal conditions under which the diversity facilitated by the primary process drive objects can enhance fitness success. Hence, the argument goes through; these primary process drive objects can be non-singular and yet have determinate content, with selective fitness normativity underwriting the constraint against indeterminancy.

2. 8. Concluding remarks: Frog representations, vague concepts, and primary process drive objects

What is the relation of the representations associated with primary process drive objects, to those frogs have regarding bugs and BBs, and to those associated with "vague concepts", a current topic in philosophy? Williamson (1994) in his tour de force book, *Vagueness*, after an exhaustive review of all competing views, essentially concludes that vague concepts result from cognitive limitations. Vague concepts pick out something precise ontologically – there is a particular number of grains of sand at which point they will constitute a "heap"; there is a specific number of hairs on his head below which he will be bald – just as there *is* one and only one number that represents the precise number of people on Main Street in Ann Arbor right now – however our cognitive limitations do not let us *know* these precise numbers. (See p. 208, in particular). A similar case could be made for frogs' representations of bugs and BBs. Although they cannot distinguish bugs from BBs, and presumably represent them identically, this representation, on the proper function view I've been propounding, always refers to bugs – correctly when there is a bug to swallow, and incorrectly when a BB is ingested, but there is a correct representation. Yes frogs' representations are good enough; but if frogs were better cognizers, they could make the distinction and perhaps there would even be some selective advantage – fewer ingestings of non-nutritive item, which even if not harmful take up the time and space that could be devoted to real bugs.

Representations of primary process amalgamated drive objects, on the other hand, bear only a superficial relation to vague concepts – one perhaps based on an insufficient understanding of the nature of the primary process mental content in general and the nature of primary process drive objects in particular. Regarding primary process content itself, consider that *intrinsically* this content might be neither vague nor indeterminate; but only seem so *relative* to the secondary process terms in which we (as secondary process dominated thinkers) are forced to formulate it. Perhaps for any representational system (including that of the frogs and their bugs/BBs) there is in itself no vagueness or indeterminacy, but only the appearance of such when there is an attempt to translate it into a *relatively* more precise and determinate representational system. While the more precise and determinate second system is capable of the distinctions that the first system cannot make, the formulation or translation may not be workable, a situation which does not imply the vagueness or indeterminacy of the first system content.

Let's turn now to the nature of primary process drive objects. Indeed as I look at the crowd at University of Michigan football stadium, to me it seems equally likely that there are 99,546 people as 90,000 people as 98,765 people. The set of numbers comprising my representation is vague; and the elements I single out when I give a number are arbitrary and random. While it may seem to be similar that an unconstrained string of random and arbitrary elements can comprise a primary process drive object, as I have tried to show, this is not the case. Moreover, the amalgam or set must not only be derived according to its proper function, following the laws of primary process toward diversity in satisfaction, but is further constrained by the need for each element to be subjectively (if idiosyncratically) relevant to the individual with the drive. Thus my mother's face will not suffice to satisfy the oral drives of Baby X; and the realm of possible Oedipal choices for me will not be the same as the realm for you. Further, improving cognitive capabilities would not seem to lead to the ability to constitute a "more correct" drive object, in the sense that without my human limitations I could more correctly know the number of hairs on Z's head below which would constitute baldness or the number of people at the U of M stadium. This owes to the fact that there is no correct or incorrect with respect to this primary process amalgam. But would fewer cognitive limitations allow "better" (if not more correct) drive objects, objects that fitted better the environment, toward greater satisfaction and ultimately enhanced fitness success? It is tempting to answer yes; but at the same time it is hard to imagine any rational matching process being as resiliently successful in providing drive and drive object fit, as are the a-rational primary process drive object amalgams, with their built in diversity.

REFERENCES

Brakel, L. (2002). Phantasy and wish: A proper function account of a-rational primary process mediated mentation. *Australasian Journal of Philosophy* 80: 1–16.
Brakel, L. (2003). "Unusual" human experiences: Kant, Freud and an associationist law. *Theoria et Historia Scientiarum* 7(2): 109–116.
Brakel, L. (2004). Der Primärvorgang: Experimentalle Belege für A-rationale Organisationsprinzipien und einen philosophischen Fall von A-rationaler Referenz. In *Psychoanalyse im Dialog der Wissenschaften. Vol. 2. Angloamerikanische Perspektiven*, ed. P. Giampieri-Deutsch. Stuttgart: Kohlhammer, pp. 143–167.
Cherniak, R. (1981). Minimal rationality. *Mind* 90: 161–183.
Davidson, D. (1980). *Actions and Events*. Oxford: Clarendon Press.
Davidson, D. (1984). *Truth and Interpretation*. Oxford: Clarendon Press.
Dennett, D. (1978). *Brainstorms*. Cambridge, Mass.: MIT Press.
Dennett, D. (1987). *The Intentional Stance*. Cambridge, Mass.: MIT Press.
Erikson, E. (1950). *Childhood and Society*. New York: Norton.
Freud, A. (1936). *The Ego and the Mechanisms of Defense*. In *The Writings of Anna Freud*. Vol. 2. New York: International Universities Press 1966.
Freud, A. (1977). A study guide to Freud's Writings. In *The Writings of Anna Freud*. Vol. 8. New York: International Universities Press 1981, pp. 209–276.
Freud, S. (1900a). *The Interpretation of Dreams*. *S.E.* 4–5.
Freud, S. (1905e). A case of hysteria. *S.E.* 7: 7–122.
Freud, S. (1905d). *Three Essays on Sexuality*. *S.E.* 7: 135–243.
Freud, S. (1909b). Analysis of a phobia in a five year old boy. *S.E.* 10: 5–147.
Freud, S. (1909d). Notes upon a case of obsessional neurosis. *S.E.* 10: 155–249.
Freud, S. (1915c). Instincts and their vicissitudes. *S.E.* 14: 117–140.
Freud, S. (1915e). The unconscious. *S.E.* 14: 166–204.
Freud, S. (1918b). An infantile neurosis. *S.E.* 17: 7–122.
Freud, S. (1926d). *Inhibitions Symptoms and Anxiety*. *S.E.* 20: 87–172.
Freud, S. (1940a). *An Outline of Psycho-Analysis*. *S.E.* 23:144–207.
Hartmann, H. (1958). *Ego psychology and the problem of adaptation*. New York: International Universities Press.
Lorenz, K. (1950). The comparative method in studying innate bird behavior patterns. In *Physiological Mechanisms in Animal Behavior, Symposia of the Society for Experimental Biology*. New York: Academic Press.
Millikan, R. (1984). *Language, Thought, and Other Biological Categories*. Cambridge, Mass.: MIT Press.
Millikan, R. (1993). *White Queen Psychology and Other Essays for Alice*. Cambridge, Mass.: MIT Press.
Nagera, H. (1969). *Basic Psychoanalytic Concepts on the Libido Theory*. Vol. 1. New York: Basic Books.
Nagera, H. (1970). *Basic Psychoanalytic Concepts on the Theory of Instincts*. Vol. 3. New York: Basic Books.
Rapaport, D. (1967). *The Collected Papers of David Rapaport*. New York: Basic Books.

Stich, S. (1983). *From Folk Psychology to Cognitive Science: The Case Against Belief.* Cambridge, Mass.: MIT Press.
Timbergen, N. (1951). *The Study of Instinct*. Oxford: Clarendon Press.
Velleman, D. (2002). Motivation by ideal. *Philosophical Explorations* 5: 89–104.
Williamson, T. (1994). *Vagueness*. London and New York: Routledge.

IV. CAUSAL INTERACTIONS OF CONSCIOUSNESS, UNCONSCIOUS MIND AND BRAIN[1]

MAX VELMANS

1. WHAT NEEDS TO BE EXPLAINED

The potential effect of the mind on the body is also taken for granted in psychosomatic medicine. But *how* the conscious mind exercises its influence is not easy to understand. In principle, there are four distinct ways in which body/brain and mind/consciousness might enter into causal relationships. There might be physical causes of physical states, physical causes of mental states, mental causes of mental states, and mental causes of physical states. Establishing which forms of causation are effective in *practice* is important, not just for a deeper understanding of mind/body interactions, but also for the proper treatment of some forms of illness and disease.

Within conventional medicine, physical→physical causation is taken for granted. Consequently, the proper treatment for physical disorders is assumed to be some form of physical intervention. Psychiatry takes the efficacy of physical→mental causation for granted, along with the assumption that the proper treatment for psychological disorders may involve psychoactive drugs, neurosurgery and so on. Many forms of psychotherapy take mental→mental causation for granted, and assume that psychological disorders can be alleviated by means of "talking cures", guided imagery, hypnosis and other forms of mental intervention. Psychosomatic medicine assumes that mental→physical causation can be effective. Consequently, under some circumstances, a physical disorder (for example, hysterical paralysis) may require a mental (psychotherapeutic) intervention. Given the extensive evidence for *all* these causal interactions (cf. readings in Velmans 1996a), how are we to make sense of them?

2. CLINICAL EVIDENCE FOR THE CAUSAL EFFICACY OF CONSCIOUS MENTAL STATES

The problems posed by mental→physical causation are particularly acute, as reductionist, materialistic science generally takes it for granted

---

[1] This chapter is based on Velmans, M. (2002a) How could conscious experiences affect brains? (A target article for a Special Issue of *Journal of Consciousness Studies*).

that the operation of physical systems can be entirely explained in physical terms. Yet there is a large body of evidence that states of mind can affect not only subsequent states of the mind but also states of the body. For example, Barber (1984), Sheikh et al. (1996), and the readings in Sheikh (2001) review evidence that the use of imagery, hypnosis, and biofeedback may be therapeutic in a variety of medical conditions.

Particularly puzzling is the evidence that under certain conditions, a range of *autonomic* body functions including heart rate, blood pressure, vasomotor activity, blood glucose levels, pupil dilation, electrodermal activity, and immune system functioning can be influenced by conscious states. In some cases these effects are striking. Baars & McGovern (1996) for example report that.

"the global influence of consciousness is dramatized by the remarkable phenomenon of biofeedback training. There is firm evidence that *any* single neurone or *any* population of neurons can come to be voluntarily controlled by giving conscious feedback of their neural firing rates. A small needle electrode in the base of the thumb can tap into a single motor unit – a muscle fibre controlled by one motor neurone coming from the spinal cord, and a sensory fibre going back to it. When the signal from the muscle fibre is amplified and played back as a click through a loudspeaker, the subject can learn to control his or her single motor unit – one among millions – in about ten minutes. Some subjects have learned to play drumrolls on their single motor units after about thirty minutes of practice! However, if the biofeedback signal is not conscious, learning does not occur. Subliminal feedback, distraction from the feedback signal, or feedback via a habituating stimulus – all these cases prevent control being acquired. Since this kind of learning only works for *conscious* biofeedback signals, it suggests again that consciousness creates global access to all parts of the nervous system." (p. 75)

The most well accepted evidence for the effect of states of mind on medical outcome is undoubtedly the "placebo effect" – well known to every medical practitioner and researcher. Simply receiving treatment, and having confidence in the therapy or therapist has itself been found to be therapeutic in many clinical situations (cf. Skrabanek & McCormick 1989; Wall 1996). As with other instances of apparent mind/body interaction, there are conflicting interpretations of the causal processes involved. For example, Skrabanek & McCormick (1989) claim that placebos can affect illness (how people feel) but not disease (organic disorders). That is, they accept the possibility of mental→mental causation but not of mental→physical causation.

However, Wall (1996) cites evidence that placebo treatments may produce organic changes. Hashish et al. (1988) for example, found that use of an impressive ultrasound machine reduced not only pain, but also jaw tightness and swelling after the extraction of wisdom teeth whether or not the machine was set to produce ultrasound. Wall also reviews evidence that placebos can remove the sensation of pain accompanying well-defined organic disorders, and not just the feelings of discomfort, anxiety and so on that may accompany it.

As McMahon and Sheikh (1989) note, the absence of an acceptable theory of mind/body interaction within philosophy and science has had a detrimental effect on the acceptance of mental causation in many areas of clinical theory and practice. Conversely, the extensive evidence for mental causation within some clinical settings forms part of the database that any adequate theory of mind/consciousness – body/brain relationships needs to explain.

3. SOME USEFUL ACCOUNTS OF MENTAL CAUSATION

The theoretical problems posed by mental causation are nicely illustrated by studies of imagery. According to the evidence reviewed by Sheikh et al (1996), imagery can be an effective tool in exercising mental control over ones own bodily states (heart rate, blood pressure, vasomotor activity and so on). It can also affect other states of mind, playing an important role in hypnosis and meditation. But, how could ephemeral images affect the spongy material of brains? And by what mechanism could conscious images affect other conscious states?

In clinical practice, the effects of imagery on brain, body and other conscious experience are often explained to patients in terms of *refocusing and redirection of attention*, linked where plausible to the operation of known biological mechanisms. For example, in their pain control induction programme, Syrjala and Abrams (1996) explain the effectiveness of imagery to patients in terms of the gate-control theory of pain:

"Even though the pain message starts in your leg, you won't feel pain unless your brain gets the pain message. The pain message moves along nerves from where the injury is located to the brain. These nerves enter the spinal cord, where they connect to other nerves, which send information up the spinal cord to the brain. The connections in the spinal cord and brain act like gates. These gates help you to not have to pay attention to all the messages in your body all the time. For example, right now as you are listening, you do not notice the feelings in your legs, although those feelings are there if you choose to notice them. If you are walking, you might notice feelings in your legs but not in your mouth. One way we block the gates to pain is with medications. Or we can block the gates by filling them with other messages. You do this if you hit your elbow and then rub it hard. The rubbing fills the gate with other messages, and you feel less pain. You've done the same thing if you ever had a headache and you get busy doing something that takes a lot of concentration. You forget about the headache because the gates are full of other messages. Imagery is one way to fill the gate. You can choose to feel the pain if you need to, but any time you like you can fill the gate with certain thoughts and images. Our goal is to find the best gate fillers for you" (p. 243).

While this account is nicely judged in terms of its practical value to patients, it does not give much detail about the actual mechanisms involved. Nor does it serve as a general account of mental causation in situations that seem to demand a more sophisticated understanding of the intricate, reciprocal balance of mind/brain/body relationships. The evidence that involuntary processes can sometimes be brought under voluntary control, for example, appears to blur the classical boundary between voluntary and autonomic nervous system functions, and extends the potential scope of top-down processing in the brain. And the evidence that imagery can sometimes have bodily effects that resemble the effects of the imaged situations themselves suggest that the conventional, clear distinction between "psychological reality" and "physical reality" may not be so clear in the way that these are *responded to* by body and brain. As Kenneth Pelletier (1993) puts it.

"Asthmatics sneeze at plastic flowers. People with a terminal illness stay alive until after a significant event, apparently willing themselves to live until a graduation ceremony, a birthday milestone, or a religious holiday. A bout of rage precipitates a sudden, fatal heart attack. Specially trained people can voluntarily control such "involuntary" bodily functions as the electrical activity of the brain, heart rate, bleeding, and even the body's response to infection. Mind and body are inextricably linked, and their second-by-second interaction exerts a profound influence upon health and illness, life and death. Attitudes, beliefs, and emotional states ranging from love and compassion to fear and anger can trigger chain reactions that affect blood chemistry, heart rate, and the activity of every cell and organ system in the body – from the stomach and gastrointestinal tract to the immune system. All of that is now indisputable fact. However, there is still great debate over the extent to which the mind can influence the body and the precise nature of that linkage" (p. 19).

One productive route to a deeper understanding of such linkages is the traditional biomedical one, involving a fuller understanding of the interconnections and reciprocal control between cortical, neuroendocrine, autonomic and immune systems. These have been extensively investigated within psychoneuroimmunology. Following a detailed review of this research, Watkins (1997) concludes that.

"It is apparent that the immune system can no longer be thought of as autoregulatory. Virtually every aspect of immune function can be modulated by the autonomic nervous system and centrally produced neuropeptides. These efferent neuroimmunomodulatory pathways are themselves modulated by afferent inputs from the immune system, the cortex and the limbic emotional centers. Thus the brain and the immune system communicate in a complex bidirectional flow of cytokines, steroids and neuropeptides, sharing information and regulating each other's function. This enables the two systems to respond in an integrated manner to environmental challenges, be they immunological or behavioral, and thereby maintain homeostatic balance" (p. 15).

4. SO WHY DOES MENTAL CAUSATION REMAIN A PROBLEM?

Such innovative findings and their practical consequences for the development of "mind-body medicine" demand careful investigation. It is important to note however that such explanatory accounts routinely translate *mind*-body interactions into *brain*-body interactions. Unless one is prepared to accept that mind and consciousness are *nothing more* than brain processes[2] this finesses the classical mind/body problems that are

[2] Although variants of eliminative/reductive physicalism and functionalism (that consciousness is nothing more than a state or function of the brain) are commonly adopted in current philosophy and science, the reduction of conscious phenomenology to brain states or functions faces well-recognised difficulties. I present a detailed

already posed by *normal* voluntary, "mental" control. How imagery might affect autonomic or immune system functioning is mysterious, but how a conscious wish to lift a finger makes that finger move is equally mysterious. Why? There are many reasons, but I will focus on just three:

Problem 1. The physical world appears causally closed. As noted above, it is widely accepted in science that the operation of physical systems can be entirely explained in physical terms. For example, if one examines the human brain from an external third-person perspective one can, in principle, trace the effects of input stimuli on the central nervous system all the way from input to output, without finding any "gaps" in the chain of causation that consciousness might fill. Indeed, the *neural correlates* of consciousness would fill any "gaps" that might potentially be filled by consciousness in the activities of brain. In any case, if one inspects the operation of the brain from the outside, no subjective experience can be observed at work. Nor does one need to appeal to the existence of subjective experience to account for the neural activity that one *can* observe. The same is true if one thinks of the brain as a functioning system described in information processing terms rather than neural terms. Once the processing within a system required to perform a given function is sufficiently well specified in procedural terms, one does not have to add an "inner conscious life" to make the system work. In principle, the same function, operating to the same specification, could be performed by a non-conscious machine.

Problem 2. One is not conscious of one's own brain/body processing. So how could there be conscious control of such processing? How "conscious" is conscious, voluntary control? It is surprising how few people

---

analysis of the strengths and weaknesses of various eliminative, reductive and emergent forms of physicalism, along with psychofunctionalism (functionalism in cognitive psychology) and computational functionalism (functionalism in philosophy and AI) in Velmans (2000) chapters 3, 4 and 5. On-line papers addressing many of the difficulties, for example in the work of Searle, Dennett, Armstrong, Block and Tye are also available from the CogPrints archive (http://cogprints.soton.ac.uk/) - see Velmans (1998, 2001a, 2001b). An extensive evaluation of physicalist solutions to the problem of how consciousness might have causal interactions with brain is also given in Velmans (2002a), eight commentaries on this target article (particularly the commentaries by Torrance, and Van Gulick) and the reply in Velmans (2002b), now jointly published as a book (Velmans 2003).

bother to ask.[3] One might be aware of the fact *that* relaxing imagery can lower heart rate, but one has no awareness of *how* it does so, nor, in biofeedback, does one have any awareness of how consciousness might control the firing of a single motor neurone. One isn't even conscious of *how* to control the articulatory system in everyday "conscious speech"! Speech production is one of the most complex tasks humans are able to perform. Yet, one has no awareness whatsoever of the motor commands issued from the central nervous system that travel down efferent fibers to innervate the muscles, nor of the complex motor programming that enables muscular coordination and control. In speech, for example, the tongue may make as many as 12 adjustments of shape per second – adjustments that need to be precisely coordinated with other rapid, dynamic changes within the articulatory system. According to Lenneberg (1967), within one minute of discourse as many as 10 to 15 thousand neuromuscular events occur. Yet only the *results* of this activity (the overt speech) normally enter consciousness.

Preconscious speech control might of course be the result of *prior* conscious activity, for example, planning *what* to say might be conscious, particularly if one is expressing some new idea, or expressing some old idea in a novel way. Speech production is commonly thought to involve hierarchically arranged, semantic, syntactic, and motor control systems in which communicative intentions are translated into overt speech in a largely top-down fashion. Planning *what* to say and translating nonverbal conceptual content into linguistic forms requires effort. But to what extent is such planning conscious? Let us see.

A number of theorists have observed that periods of conceptual, semantic and syntactic planning are characterized by gaps in the otherwise relatively continuous stream of speech (Goldman-Eisler 1968; Boomer 1970). The neurologist John Hughlings Jackson, for example, suggested that the amount of planning required depends on whether the speech is "new" speech or "old" speech. Old speech (well-known phrases, etc.) requires little planning and is relatively continuous. New speech (saying things in a new way) requires planning and is characterized by hesitation pauses. Fodor, Bever & Garrett (1974) point out that breathing pauses also occur (gaps in the speech stream caused by the intake of breath). However, breathing pauses do not generally coincide with hesitation pauses.

Breathing pauses nearly always occur at the beginnings and ends of major linguistic constituents (such as clauses and sentences). So these appear to be coordinated with the syntactic organization of such constitu-

[3] See the initial discussion of this issue in Velmans (1991a).

ents into a clausal or sentential structure. Such organization is largely automatic and preconscious. By contrast, hesitation pauses tend to occur within clauses and sentences and appear to be associated with the formulation of ideas, deciding which words best express one's meaning, and so on. If this analysis is correct, conscious planning of *what* to say should be evident during hesitation pauses – and a little examination of what one experiences during a hesitation pause should settle the matter. Try it. During a hesitation pause one might experience a certain sense of effort (perhaps the effort to put something in an appropriate way). But nothing is revealed of the *processes* that formulate ideas, translate these into a form suitable for expression in language, search for and retrieve words from memory, or assess which words are most appropriate. In short, no more is revealed of conceptual or semantic planning in hesitation pauses than is revealed of syntactic planning in breathing pauses. The fact that a process demands processing *effort* does not ensure that it is *conscious*. Indeed, there is a sense in which one is only conscious of what one wants to say *after one has said it*!

It is particularly surprising that the same may be said of *conscious verbal thoughts*. That is, the same situation applies if one formulates one's thoughts into "covert speech" through the use of phonemic imagery, prior to its overt expression. Once one *has* a conscious verbal thought, manifested in experience in the form of phonemic imagery, the complex cognitive processes required to generate that thought, including the processing required to encode it into phonemic imagery *have already operated*. In short, covert speech and overt speech have a similar relation to the planning processes that produce them. In neither case are the complex antecedent processes available to introspection. It should be clear that this applies equally to the processes that generate the detailed spatial arrangement, colours, shapes, sizes, movements and accompanying sounds and smells of an imaged visual scene.

Problem 3. Conscious experiences appear to come too late to causally affect the processes to which they most obviously relate. In the production of overt speech and covert speech (verbal thoughts) the conscious experience that we normally associate with such processing *follows* the processing to which it relates. Given this, in what *sense* are these "conscious processes" conscious? The same question can be asked of that most basic of conscious voluntary processes, *conscious volition itself.*

It has been known for some time that voluntary acts are preceded by a slow negative shift in electrical potential (recorded at the scalp) known as the "readiness potential", and that this shift can precede the act by up to one second or more (Kornhuber & Deeke 1965). In itself, this says nothing about the relation of the readiness potential to the *experienced wish* to perform an act. To address this, Libet (1985) asked subjects to note the instant they experienced a wish to perform a specified act (a simple flexion of the wrist or fingers) by relating the onset of the experienced wish to the spatial position of a revolving spot on a cathode ray oscilloscope, which swept the periphery of the face like the sweep-second hand of a clock. Recorded in this way, the readiness potential preceded the voluntary act by around 550 milliseconds, and preceded the experienced wish (to flex the wrist or fingers) by around 350 milliseconds (for spontaneous acts involving no preplanning). This suggests that, like the act itself, the experienced wish (to flex one's wrist) may be one output from the (prior) cerebral processes that actually select a given response. If so, "conscious volition" may be no more necessary for such a (preconscious) choice than the consciousness of ones own speech is necessary for its production.[4] And the same is likely to apply to more complex voluntary acts, such as the voluntary control of autonomic functions through imagery and biofeedback discussed above.

5. WHAT IS A CONSCIOUS MENTAL PROCESS?

It should be clear from the evidence above that we need to distinguish three different senses in which a mental process might be said to

[4] As Libet observed, the experienced wish *follows* the readiness potential, but *precedes* the motor act itself (by around 200 msec) - time enough to consciously *veto* the wish before executing the act. In a manner reminiscent of the interplay between the libidinous desires arising from Freud's unconscious *id* and the control exercised by the conscious *ego*, Libet suggested that the *initiation* of voluntary act and the accompanying wish are developed preconsciously, but consciousness can then act as a form of censor which decides whether or not to carry out the act. While this is an interesting possibility, it does invite an obvious question. If the wish to perform an act is developed preconsciously, why doesn't the decision to censor the act have its own preconscious antecedents? Libet (1996) argues that it *might* not need to do so as voluntary control imposes a change on a wish that is already conscious. Yet, it seems very odd that a wish to do something has preconscious antecedents while a wish not to do something does not. As it happens, there is evidence that bears directly on this issue. Karrer, Warren & Ruth (1978), and Konttinen & Lyytinen (1993), for example, found that *refraining* from irrelevant movements is associated with a slow *positive-going* readiness potential.

"be conscious" (cf. Velmans 1991). A mental process might be said to be conscious (a) in the sense that we are conscious *of* it, (b) in that it *results* in a conscious experience, and (c) in that conscious phenomenology causally affects or enters into that process. We do not have introspective access to how the preconscious cognitive processes that enable thinking produce individual, conscious thoughts in the form of "inner speech." However, the content of such thoughts and the sequence in which they appear does give some insight into the way the cognitive processes (of which they are manifestations) operate over time in problem solving, thinking, planning and so on. Consequently such cognitive processes are partly conscious in sense (a), but only in so far as their detailed operation is made explicit in conscious thoughts, thereby becoming accessible to introspection. Many psychological processes are conscious in sense (b), but not in sense (a) – that is, we are not conscious of how they operate, but we are conscious of their *results*. This applies to perception in all sense modalities. When consciously reading this sentence for example you become aware of the printed text on the page, accompanied, perhaps, by inner speech (phonemic imagery) and a feeling of understanding (or not), but you have no introspective access to the processes which enable you to read. Nor does one have introspective access to the *details* of most other forms of cognitive functioning, for example to the detailed operations which enable "conscious" learning, remembering, engaging in conversations with others and so on.

Crucially, having an experience that gives some introspective access to a given process, or having the results of that process manifest in an experience, says nothing about whether that experience *carries out* that process. That is, whether a process is "conscious" in sense (a) or (b) needs to distinguished from whether it is conscious in sense (c). Indeed, it is not easy to envisage how the experience that makes a process conscious in sense (a) or (b), *could* make it conscious in sense (c). Consciousness *of* a physical process does not make consciousness responsible for the operation of that process (watching a kettle does not determine when it comes to the boil). So, how could consciousness *of* a mental process carry out the functions of that process? Alternatively, if conscious experience *results* from a mental process it arrives *too late* to carry out the functions of that process (see Velmans 2000, chapter 9 for a more detailed discussion).

Over 100 years ago, Freud expressed similar views about the relationship of conscious experience to the operations of preconscious mental processing and about the limited causal role of consciousness:

"... the unconscious must be assumed to be the primary basis of psychic life ... Everything conscious has an unconscious preliminary stage; whereas what is unconscious may remain at that stage and nevertheless claim to be regarded as having the full value of a psychical process. The unconscious is the true psychical reality; *in its innermost nature it is as much unknown to us as the reality of the external world, and it is as incompletely presented by the data of consciousness as is the external world by the communications of our sense organs*" (Freud 1900a, pp. 612–613).

At the same time, in much of Freud's work, it is assumed that the functions of the *ego*, operating according to the *reality principle*, are conscious. Consequently, the causal interaction between conscious and unconscious mind remains a central issue for psychoanalytic theory just as it is for consciousness studies. The same is true of the interactions of mind and body/brain. From studies of psychosomatic effects such as hysterical paralysis, Freud had no doubts that mind/body interactions are real. As he noted "a disturbance that can be set right by means of psychic influences can itself have been nothing else than psychical" (quoted in Ferenczi, et al. 1921, p. 11). At the same time he was pessimistic about ever finding a satisfactory explanation. In his view, there is a leap from "a mental to a somatic innervation" (e.g. in hysterical paralysis) that "can never be fully comprehensible to us" (Freud 1909d, p. 157).

6. THE CURRENT THEORETICAL IMPASSE

As noted, there is extensive experimental and clinical evidence that conscious experiences can affect brain/body processes, and the importance of conscious experience is rightly taken for granted in everyday life. In one sense this can be explained by a more sophisticated biomedical understanding of mind/brain/body relationships. But in a deeper sense, current attempts to understand the role of conscious experience face an impasse. How can experiences have a causal influence on a physical world that is causally closed? How can experiences affect processes that *precede* them? And how can one consciously control processes that are basically unconscious? Dualist-interactionist accounts of the consciousness-brain relationship, in which an autonomously existing consciousness influences the brain, do not even recognise these "how" problems let alone address them. Materialist reductionists attempt to finesse such problems by challenging the accuracy, causal efficacy and even the existence of conscious experiences. This evades the need to address the "how" questions, but denies the validity of the clinical evidence and defies common sense. I have given a detailed critique of the many variants of dualism and reductionism elsewhere and will not repeat this here.[5] In what follows I suggest a way through the impasse that is neither dualist nor reductionist.[6]

7. ONTOLOGICAL MONISM COMBINED WITH EPISTEMOLOGICAL DUALISM

How can one reconcile the evidence that conscious experiences are causally effective with the principle that the physical world is causally closed? One simple way is to accept that for each individual there is *one* "mental life" but *two* ways of knowing it: first-person knowledge and third-person knowledge. From a first-person perspective conscious experiences appear causally effective. From a third-person perspective the same causal sequences can be explained in neural terms. It is not the case that the view from one perspective is right and the other wrong. These perspectives are complementary. The differences between how things appear from a first versus a third person perspective have to do with differences in the *observational arrangements* (the means by which a subject and an external observer access the subject's mental processes).

Let's see how this might work in practice. Suppose you have a calming image of lying in a green field on a summer's day, and you can feel the difference this makes in producing a relaxed state, slowing your breathing, removing the tension in your body and so on. You give a causal account of what is going on, based on what you experience. From my external observer's perspective, I can also observe what is going on – but what I observe is a little different. I can measure the effects on your breathing and muscle tension, but no matter how closely I inspect your brain, I cannot observe your experienced image. The closest I can get to it are its neural correlates in the visual system, association areas and so on.[7] Nevertheless, if I could observe all the neurophysiological

[5] See references in note 2.

[6] In the space available I can give only an introduction to how one might resolve these problems. A more detailed treatment is given in Velmans (2000) chapter 11 and Velmans (2003).

[7] The neural correlates of a given experience accompany or *co-occur* with given experiences, and are by definition as close as one get to those experiences from an

events operating in your brain to produce your relaxed bodily state, I could give a complete, physical account of what is going on. So, now you have a first-person account of what is going on that makes sense to you and I have a third-person account of what is going on that makes sense to me. How do these relate? To understand this we need to examine the relation of your visual image to its neural correlates with care.

8. THE NEURAL CORRELATES OF CONSCIOUS EXPERIENCE

Although we know little about the physical nature of the neural correlates of conscious experiences, there are three plausible, functional constraints imposed by the phenomenology of consciousness itself. Normal human conscious experiences are representational (phenomenal consciousness is always *of* something).[8] Given this, it is reasonable to as-

external observer's perspective. This differentiates them from the antecedent causes (such as the operation of selective attention, binding, etc.) which may be thought of as the necessary and sufficient *prior* conditions for given experiences in the human brain.

[8] My assumption that normal conscious experiences are representational is driven by a Critical Realist epistemology (developed in Velmans, 2000, chapter 7) and not by any commitment to the view that mental states are nothing more than computations on representations (a thesis that is currently in dispute). While I do not have space to develop the case for Critical Realism here, it is worth noting that there is nothing mysterious about experiences being representations of entities and events outside of or within our bodies and brains that differ in some respects from the alternative representations of those entities and events given by science (e.g. by physics). Perceptual processes are likely to have developed in response to evolutionary pressures, and select, attend to, and interpret information in accordance with human adaptive needs. Consequently, they only need to model a subset of the available information. At the same time our perceptual models must be useful, otherwise it is unlikely that human beings would have survived. Given this, it seems reasonable to assume that, barring illusions or hallucinations, the experiences produced by perceptual processing are partial, approximate but nonetheless useful representation of what is "really there". The view that some conscious experiences are representational in the sense of being "intentional" (that they are *of* something) has in any case been widely accepted in philosophy of mind since Brentano reintroduced this medieval notion in the 19th Century. According to some philosophers not all conscious experiences are intentional. Searle (1994) for example maintains that "a feeling of pain or a sudden sense of anxiety, where there is no object of the anxiety, are not intentional" (p. 380). In Velmans (1990, 2000) I argue that a conscious experience does not have to be about a specific external object for it to be representational. It may for example represent a state of one's own body or it may represent a global *reaction* to a real, imagined or

sume that the neural correlates of such experiences are also representational states.

Although this assumption has not always been made explicit in theories of consciousness it is largely taken for granted in psychological theory. Psychophysics, for example, takes it for granted that for any discriminable aspect of experiences (a just noticeable change in brightness, colour, pitch and so on) there will be a correlated change in some state of the brain. It follows from this that the information encoded in experience (in terms of discriminable differences) will also be encoded in the brain. The same is true for the more complex contents of consciousness, in the many cognitive theories that associate (or identify) such contents with information stored in primary (working) memory, information at the focus of attention, information in a global workspace and so on.

A representational state must, of course, represent *something*, that is it must have a given content. For a given physical state to be the correlate of a given experience it is plausible to assume that it represents the *same* thing (otherwise it would not be the correlate of *that* experience).

Finally, for a physical state to be the correlate of a given experience, it is reasonable to suppose that it has the same "grain". That is, for every discriminable attribute of experience there will be a distinct, correlated, physical state. As each experience and its physical correlate represents the same thing it follows that each experience and its physical correlate encodes the same information about that thing. That is, they are representations with the same *information structure*.[9] [10]

---

remembered event. A feeling of pain, for example, represents (in one's first person experience) actual or potential damage to the body, and it is usually quite accurate in that it is normally subjectively located at or near the site of body damage. A feeling of anxiety is a first-person representation of a state of ones own body and brain that signals actual or potential danger, and so on. Viewed this way, *all* conscious states are about something. On this issue, I adopt the same stance as that developed by Tye (1995) - and a similar view appears to have been adopted by Freud (see quotation from *The Interpretation of Dreams*, above).

[9] This assumption of conscious experience/ neural correlate functional equivalence (defined in information processing terms) is a point of convergence between otherwise widely divergent theories (physicalism, functionalism, dual-aspect theory). As Gardner (1987) points out, the assumption that mental processes operate on representations lies at the foundations of cognitive science. However, the claim that the neural correlates of conscious states are representations begs no questions about the forms that these representations might take, or about how mental processes operate on them. Representations might be iconic, propositional, feature sets, proto-

If these assumptions are well founded, your experience and the neural correlates that I observe will relate to each other in a very precise way. What you experience takes the form of visual or other imagery accompanied by feelings about lying on the grass on a summery day. What I observe *is the same information* (about the visual scene) encoded in the physical correlates of what you experience in your brain. The information structure of what you and I observe is identical, although it is displayed or "formatted" in very different ways. From your point of view, the only information you have about your own state of mind is the imagery and accompanying feelings that you experience. From my point of view, the only information you have (about your own state of mind) is the information I can see encoded in your brain. The way your information (about your own state) is displayed appears to be very different to you and me for the reason that the "observational arrangements" by which we access that information are entirely different. From my external, third person perspective I can only access the information encoded in your mind/brain by means of my visual or other exteroceptive systems aided by appropriate equipment. With these means I can detect the information displayed in the form of neural encodings, but not in the form of accompanying experiences. While you maintain your focus on the imaged scene, you cannot observe its neural correlates in your own brain (you would need to use my equipment for that). Nevertheless, the

---

types, procedural, localised, distributed, static or dynamic, or whatever. Operations might be formal and computational, or more like the patterns of shifting weights and probabilities that determine the activation patterns in neural networks. I suggest that the correlates of consciousness represent what the phenomenology itself represents, irrespective of how the correlates embody those representations.

[10] This approach has its origins in Spinoza's dual-aspect theory, which I developed into a naturalised, dual-aspect theory of information in Velmans (1991a, 1991b, 1993, 1996, 2000). This dual-aspect theory of information also has similarities to that adopted by Chalmers (1996) (see Velmans 2000, p. 281, note 5 for a summary of both the similarities and differences). Note that having an identical referent and information structure does not mean that experiences are *nothing more than* their neural correlates (as eliminativists and reductionists assume). A filmed version of the play "Hamlet", recorded on videotape, for example, may have the same sequential information structure as the same play displayed in the form of successive, moving pictures on a TV screen. But it is obvious that the information on the videotape is not ontologically identical to the information displayed on the screen. In this instance, the same information is embodied in two different ways (patterns of magnetic variation on tape versus patterns of brightness and hue in individual pixels on screen) and it is displayed or "formatted" in two different ways (only the latter display is in visible form).

information in those correlates displays "naturally"[11], in the form of the imaged scene that you experience.

But what is your mind *really* like? From my "external observer's perspective", can I assume that what you experience is really nothing more than the physical correlates that I can observe? From my external perspective, do I know what is going on in your mind/brain/consciousness better than you do? No. I know something about your mental states that you do not know (their physical embodiment). But you know something about them that I do not know (their manifestation in your experience). Such first- and third-person information is *complementary*. We need your first-person story and my third-person story for a complete account of what is going on. If so, the nature of the mind is revealed as much by how it appears from one perspective as the other. It is not *either* physical *or* conscious experience, it is at once physical *and* conscious experience (depending on the observational arrangements). For lack of a better term we may describe this nature as *psychophysical*.[12] [13] If we combine this with the representational features

[11] I assume it to be a *natural* fact about the world that certain forms of neural activity are accompanied by conscious experiences. Consequently, when such neural activities (the correlates) occur in one's brain one has the corresponding experiences. I also assume that the formatting of neurally encoded information relates to the formatting of corresponding, phenomenally encoded information in an orderly way, with discoverable neural state space/phenomenal space mappings. An obvious example would be the way that information about spatial location and extension encoded in the brain is mapped into the 3D phenomenal space that we ordinarily experience. In vision, some progress has already been made in the discovery of such mappings (see the Special Issue on the work of Roger Shepard in *Behavioural and Brain Sciences* 24(4), 2001). While neural state/phenomenal state mappings are likely to differ in different sense modalities (e.g. vision versus audition) and even between different features of a given modality (e.g. colour versus spatial location and extension) there may also be shared, underlying principles (cf. Stoffregen & Benoît 2001).

[12] The struggle to find a model or even a form of words that somehow captures the dual-aspect nature of mind is reminiscent for example of wave-particle complementarity in quantum mechanics - although this analogy is far from exact. Light either appears to behave as electromagnetic waves or as photon particles depending on the observation arrangements. And it does not make sense to claim that electromagnetic waves really *are* particles (or vice versa). A complete understanding of light requires both complementary descriptions - with consequent struggles to find an appropriate way of characterizing the nature of light and other QM phenomena which encompass both descriptions ("wave-packets", "electron clouds" and so on). This has not prevented physics from developing very precise accounts of light viewed *either* as waves *or* as particles, together with precise formulae for relating wave-like properties

above, we can say that mind is a psychophysical process that encodes information, developing over time.

9. AN INITIAL WAY TO MAKE SENSE OF THE CAUSAL INTERACTIONS BETWEEN CONSCIOUSNESS AND BRAIN

This brief analysis of how first- and third-person accounts relate to each other can be used to make sense of the different *forms* of causal interaction that are taken for granted in everyday life or suggested in the clinical and scientific literature. Physical→physical causal accounts describe events from an entirely third-person perspective (they are "pure third-person accounts"). Mental→mental causal accounts describe events entirely from a first-person perspective (they are "pure first-person accounts"). Physical→mental and mental→physical causal accounts are *mixed-perspective* accounts employing *perspectival switching* (Velmans 1996b). Such accounts start with a description of causes viewed from one

---

(such as electromagnetic frequency) to particle-like ones (such as photon energy). If first- and third person accounts of consciousness and its physical correlates are complementary and mutually irreducible, an analogous "psychological complementarity principle" might be required to understand the nature of mind. A more detailed discussion of how psychological complementarity relates to physical complementarity is given in Velmans (2000, ch. 11, note 19).

[13] At the macrocosmic level, the relation of electricity to magnetism also provides a clear parallel to the form of dual-aspect theory I have in mind. If one moves a wire through a magnetic field this produces an electrical current in the wire. Conversely, if one passes an electrical current through a wire this produces a surrounding magnetic field. But it does not make sense to suggest that the current in the wire is nothing more than the surrounding magnetic field, or vice-versa (reductionism). Nor is it accurate to suggest that electricity and magnetism are energies of entirely different kinds that happen to interact (dualist-interactionism). Rather these are two manifestations (or "dual-aspects") of *electromagnetism*, a more fundamental energy that grounds and unifies both, described with elegance by Maxwell's Laws. Analogously, phenomenally encoded information and its correlated neurally encoded information may be two manifestations (or "dual-aspects") of a more fundamental, psychophysical mind, and their relationship may, in time, be describable by neurophenomenological laws (see also note 12 above). It goes without saying that a *fully* satisfying psychophysical account of any given mental state would have to specify how given complementary first- and third-person descriptions relate to each with precision (perhaps with the elegance of Maxwell's Laws). However, such empirical relationships can only be discovered by neuropsychological research, and for the present I am only concerned with the *form* that causal accounts based on such research might need to take to resolve this aspect of the "causal paradox".

perspective (either first- or third-person) and then switch to a description of effects viewed from the other perspective. To understand such accounts, one first has to acknowledge that a perspectival switch has taken place.

Physical→mental causal accounts start with events viewed from a third-person perspective and switch to how things appear from a first-person perspective. For example, a causal account of visual perception starts with a third-person description of the physical stimulus and the visual system but then switches to a first-person account of what the subject experiences. Mental→physical causal accounts switch the other way. From your subjective point of view, for example, the imagery that you experience is causing your heart rate to slow down and your body to relax (effects that I can measure). If I could identify the exact neural correlates of what you experience, it might be possible for me to give an entirely third-person account of this sequence of events (in terms of higher order neural representations having top-down effects on other brain and body states). But the mixed-perspective account actually gives you a more immediately useful description of what is going on in terms of the things that you can do (maintain that state of mind, deepen it, alter it, and so on).

In principle, complementary first- and third-person sources of information can be found whenever body or mind/brain states are represented in some way in subjective experience. A patient might for example have insight into the nature of a psychological problem (via feelings and thoughts) that a clinician might investigate by observing his/her brain or behaviour. In medical diagnosis, a patient might have access to some malfunction via interoceptors, producing symptoms such as pain and discomfort, whereas a doctor might be able to identify the cause via his/her exteroceptors (eyes, ears and so on) supplemented by medical instrumentation. As with conscious states and their neural correlates the clinician has access to the physical embodiment of such conditions, while the patient has access to how such conditions are experienced. In these situations, neither the third-person information available to the clinician nor the first-person information available to the patient is *automatically* privileged or "objective" in the sense of being "observer-free." The clinician merely reports what he/she observes or infers about what is going on (using available means) and the patient does likewise. Such first- and third person accounts of the subject's mental life or body states are complementary, and mutually irreducible. *Taken together*, they provide a global, psychophysical picture of the condition under scrutiny.

10. CONSCIOUS EXPERIENCES ARE CURRENT, GLOBAL REPRESENTATIONS FORMED BY THE MIND/BRAIN

The above, I hope, gives an initial indication of how one can reconcile the evidence that conscious experiences appear causally effective with the principle that the physical world is causally closed. But there are two further, equally perplexing problems. How can conscious experiences be causally effective if they come too late to affect the mind/brain processes to which they most obviously relate? And how can the contents of consciousness affect brain and body states when one is not conscious of the biological processes that govern those states?

I suggest that to make sense of these puzzles, one has to begin by accepting the facts rather than sweeping them under some obscuring theoretical carpet. Why do experiences come too late to affect the mind/brain processes to which they most closely relate? For the simple reason that experiences relate most closely to the processes that *produce* them. Visual perception becomes "conscious" once visual processing results in a conscious visual experience, cognitive processing becomes "conscious" once it produces the inner speech that forms a conscious thought and so on. Once such experiences arise the processes that have produced them have already taken place. Given this, what is consciousness actually contributing to conscious perception, to conscious speech, to conscious thought, to conscious voluntary control, and so on?

As noted above, I am proceeding on the assumption that conscious experiences are representations. Some experiences represent states of the external world (exteroceptive experiences), some represent states of the body (interoceptive experiences), and some represent states of the mind/brain itself (volitions, thoughts about thoughts, etc.). Experiences can also represent past, future, real and imaginary events, for example in the form of thoughts and images.

Whatever their representational content, current experiences also tell one something important about the current state of one's own mind/brain – that it currently has percepts, feelings, thoughts, images, etc., of a given type, and that it has formed current representations with that particular content, as opposed to any others. For example, the thoughts that enter consciousness at a given moment "represent" the current state of one's own cognitive system in that they reveal *which* of many possible cognitions are currently at the focus of attention in a reportable form. If your thoughts are conscious, and I ask you what you are thinking about, you can tell me. Likewise, your visually imaged peaceful world and your conscious feelings about it represent a current, voluntarily produced representational state (and affective responses to it) within your own visual, cognitive and affective systems – and if I want to know what that is like, you can tell me.

Why don't we have more detailed experiences of the processes that produce such conscious experiences, or of the detailed workings of our own bodies, minds and brains? Because for normal purposes we don't need them! Our primary need is to interact successfully with the external world and with each other – and for that, the processes by which we arrive at representations of ourselves in the world, or which govern the many internal, adaptive adjustments we have to make are best left on "automatic". This is exemplified by the well-accepted transition of skills from being conscious to being nonconscious as they become well learnt (as in reading or driving a car). The global representations that we have of ourselves in the world nevertheless provide a useful, reasonable accurate representation of what is going on.[14]

11. HOW TO MAKE SENSE OF THE CAUSAL ROLE OF THE CONTENTS OF CONSCIOUSNESS

As noted above, normal experiences are *of* something i.e., they represent entities, events and processes in the external world, the body and the mind/brain itself. In everyday life, we also behave as "naïve realists". That is we take the events we experience to *be* the events that are actually taking place, although sciences such as physics, biology and psychology might represent the same events in very different ways. For everyday purposes, the assumption that the world just *is* as we experience it to be serves us well. When playing billiards, for example, it is safe to assume that the balls are smooth, spherical, coloured, and cause each other to move by mechanical impact. One only has to judge the precise angle at which the white ball hits the red ball to pocket the red. A quantum mechanical description of the microstructure of the balls or of the forces they exert on each other won't improve one's game.

[14] It is reasonable to suppose that the detail of conscious representation has been tailored by evolutionary pressures to be useful for everyday human activities (although these remain global, approximate and species-specific). To obtain a more intricate knowledge of the external world, body or mind/brain we usually need the assistance of scientific instruments. A much fuller analysis of these points is given in Velmans (2000) chapter 7.

That said, the experienced world is not the world *in itself* – and it is not our experience *of* the balls that governs the movement of the balls themselves. Balls as-experienced and their perceived interactions are global *representations* of autonomously existing entities and their interactions, and conscious representations of such entities or events can only be formed once they exist, or after they have taken place. The same may be said of the events and processes that we experience to occur in our own bodies or minds/brains. When we withdraw a hand quickly from a hot iron, we experience the pain (in the hand) to cause what we do, but the reflex action actually takes place before the experience of pain has time to form. This can also happen with voluntary movements. Suppose, for example, that you are required to press a button as soon as you feel a tactile stimulus applied to your skin. A typical reaction time is 100 ms or so. It takes only a few milliseconds for the skin stimulus to reach the cortical surface, but Libet, et al. (1979) found that awareness of the stimulus takes at least 200 ms to develop. If so, the reaction must take place preconsciously, although we *experience* ourselves as responding *after* we feel something touching the skin. The mind/brain requires time to form a conscious representation of a pain or of something touching the skin and of the subsequent response. Although the conscious representations accurately place the cause (the stimulus) before the effect (the response), once the representations are formed, both the stimulus and the response have already taken place.[15]

Just as the interactions amongst experienced billiard balls represent causal sequences in the external world, but are not the events themselves, experienced interactions between our sensations, thoughts, images and actions represent causal sequences within our bodies and brains, but are not

[15] Although conscious experiences arise too late to play a causal role in the processes with which they are most closely associated (those that produce them), once they arise, they are not, of course, too late to play a causal role in other, *subsequent* mind/brain/body states or activities. A pain in the tooth, for example, might persist long enough to force one to the dentist. A desire for employment might lead to make a job application and so on. However such forms of mental→physical causation still face the problem (already discussed) that the physical world is causally closed. For example, the physical movements that take one to the dentist can be explained by the way that the neural correlates of the pain enter into the control of motor systems, the desire for employment in terms of a goal state that is represented in one's CNS and so on. Such forms of mental causation can, however, be understood as "mixed-perspective" causal accounts of the kind described above. See also the extensive treatment of this particular issue in discussion with Rakover in Velmans (1996b).

the events themselves. The thoughts, images, and feelings that appear in our awareness are both *generated by* processes in our bodies and mind/brains and *represent* the current states of those processes. Thoughts and images represent the ongoing state of play of our cognitive systems; feelings represent our internal (positive and negative) reactions to and judgements about events (see Mangan 1993, and the discussion above).

In sum, conscious representations of inner, body and external events are not the events themselves, but they generally represent those events and their causal interactions sufficiently well to allow a fairly accurate understanding of what is happening in our lives. Although they are only *representations* of events and their causal interactions, for everyday purposes we can take them to *be* those events and their causal interactions. When we play billiards we can line up a shot without the assistance of physics. Although our knowledge of our own inner states is not incorrigible, when we experience our verbal thoughts expressed in covert or overt speech, we usually know all we need to know about what we currently think – without the assistance of cognitive psychology. When we experience ourselves to have acted out of love or fear, we usually have an adequate understanding of our motivation – although a neuropsychologist might find it useful to give a third-person account of emotion in terms of its neural substrates in the neocortical, subcortical, diencephalic, midbrain and pontine-medullary brainstem systems (Watt 2000). And when we image ourselves in green grass on a summer's day and feel relaxed we are usually right to assume that the mental state that is represented in our imagery has produced a real bodily effect. For everyday life, it doesn't matter that we don't understand how such imaged scenarios are constructed by preconscious mental processes or exercise top-down control in the mind/brain/body system. It is not the case that a lower level (microscopic) representation is always better than a macroscopic one (in the case of billiard balls). Nor are third-person accounts always better than first-person ones (in describing or attempting to control our thoughts, images and emotions). The value of a given representation, description or explanation can only be assessed in the light of the purposes for which it is to be used.

12. WHO'S IN CONTROL?

The difference between voluntary and involuntary bodily functions is accepted wisdom, enshrined in the voluntary/autonomic nervous system distinction in medical texts. As we have seen above, some processes that are normally involuntary can also become partly voluntary once they are represented in consciousness (via biofeedback, imagery and so on). But if we don't have a detailed conscious awareness of the workings of our own bodies and brains and if consciousness comes too late to affect the processes to which it most closely relates how can this be? Consider again the dilemma posed by Libet et al's (1979) experiments on the role of conscious volition described above. If the brain prepares to carry out a given action around 350 milliseconds before the conscious wish to act appears, then how could that action be "conscious" and how could it be "voluntary"? Doesn't the preceding readiness potential indicate that the action is determined preconsciously and automatically by processing in the mind/brain?

Let us consider the "conscious" aspect first. The decision to act (indexed by the readiness potential) is taken preconsciously but it becomes conscious at the moment that it manifests *as* a wish to do something in conscious experience. The wish then becomes conscious in the same way that your perception of this WORD is conscious. Like the wish, once you become conscious of this WORD, the physical, syntactic and semantic analyses required to recognise it have already taken place. Nonetheless, once you become conscious of the wish or the WORD the mental/brain processes make a transition from a preconscious to a conscious state – and it is only when this happens that you consciously realise what is going on.[16]

But how could an act that is executed *preconsciously* be "voluntary"? Voluntary actions imply the possibility of choice, albeit choice based on available external and internal information, current needs and goals. Voluntary actions are also potentially flexible and capable of being novel. In the psychological literature these properties are traditionally associated with controlled rather than automatic processing or with focal-attentive rather than pre-attentive or non-attended processing.[17] Unlike automatic or pre-attentive processing, both controlled processing (in the execution of acts) and focal-attentive processing (in the analysis of input) are thought to be "conscious". None of the above ar-

[16] I do not have space to develop this theme in more detail here. In Velmans (2000) chapters 10, 11 and 12 I develop a broader "reflexive monist" philosophy in which the function of consciousness is to "real-ize" the world. That is, once an entity, event or process enters consciousness it becomes *subjectively real*.

[17] Such functional differences are beyond the scope of this paper. However they have been extensively investigated, e.g. in studies of selective attention, controlled versus automatic processing, and so on (see e.g. Velmans, 1991, Kihlstrom, 1996).

gues against such traditional wisdom. In Libet's experiments the conscious experience appears around 350 milliseconds after the onset of preconscious processes that are indexed by the readiness potential. This says something about the timing of the conscious experience in relation to the processes that generate it and about its restricted role once it appears. However, it does not argue against the voluntary nature of that preconscious processing. On the contrary, the fact that the act consciously feels as if it is voluntary and controlled suggests that the processes which have generated that experience *are* voluntary and controlled, as conscious experiences generally provide reasonably accurate representations of what is going on (see above). This applies equally to the voluntary nature of more complex, mental processing such as the self-regulating, self-modifying operations of our own psychophysical minds evidenced by the effects of conscious imagery, meditation and biofeedback. In short, I suggest that the feeling that we are free to choose or to exercise control is compatible with the nature of what is actually taking place in our own central nervous system, following processes that select amongst available options, in accordance with current needs, goals, available strategies, calculations of likely consequences and so on. While I assume that such processes operate according to determinate physical principles, the system architecture that embodies them enables the ability to exercise the choice, flexibility and control that we experience – a form of biological determinism that is compatible with experienced free will.

So *who*'s in control? Who chooses, has thoughts, generates images and so on? We habitually think of ourselves as being our *conscious selves*. But it should be clear from the above that the different facets of our experienced, conscious selves are generated by and represent aspects of our own preconscious minds. That is, we are *both* the pre-conscious generating processes *and* the conscious results. Viewed from a third-person perspective our own preconscious mental processes look like neurochemical and associated physical activities in our brains. Viewed introspectively, from a first-person perspective, our preconscious mind seems like a personal, but 'empty space' from which thoughts, images, and feelings spontaneously arise. *We* are as much one thing as the other – and this requires a shift in our sensed "centre of gravity" to one where our consciously experienced self becomes just the visible "tip" of our own embedding, preconscious mind. I suspect that Freud would wholeheartedly agree.

REFERENCES

Baars, B.J. & McGovern, K. (1996). Cognitive views of consciousness: What are the facts? How can we explain them? In *The Science of Consciousness: Psychological, Neuropsychological, and Clinical Reviews*, ed. M. Velmans. London: Routledge, pp. 63–95.

Barber, T.X. (1984). Changing "unchangeable" bodily processes by (hypnotic) suggestions: A new look at hypnosis, cognitions, imagining, and the mind-body problem. In *Imagination and Healing*, ed. A.A. Sheikh. Farmingdale, N.Y.: Bayworld.

Behavioural and Brain Sciences (2001). Special issue on the work of Roger Shepard. *Behavioral and Brain Sciences* 24(4).

Boomer, D.S. (1970). Review of F. Goldman-Eisler *Psycholinguistics: Experiments in Spontaneous Speech. Lingua* 25: 152–164.

Ferenczi, S., Abraham, K., Simmel, E. & Jones, E. (1921). *Psychoanalysis and the War Neurosis.* London: The International Psycho-Analytic Press.

Freud, S. (1900a). *The Interpretation of Dreams. S.E.* 4–5.

Freud, S. (1909d). Notes upon a case of obsessional neurosis. *S.E.* 10: 155–249.

Fodor, J.A., Bever, T.G. & Garrett, M.F. (1974). *The Psychology of Language.* New York: McGraw-Hill.

Gardner, H. (1987). *The Mind's New Science.* New York: Basic Books, Inc.

Goldman-Eisler, F. (1968). *Psycholinguistics: Experiments in Spontaneous Speech.* New York: Academic Press.

Hashish, I., Finman, C. & Harvey, W. (1988). Reduction of postoperative pain and swelling by ultrasound: A placebo effect. *Pain* 83: 303–311.

Kanttinen, N. & Lyytinen, H. (1993). Brain slow waves preceding time-locked visuo-motor performance. *Journal of Sport Sciences* 11: 257–266.

Karrer, R., Warren, C. & Ruth, R. (1978). Slow potentials of the brain preceding cued and non-cued movement: Effects of development and retardation. In *Multidisciplinary Perspectives in Event-Related Potential Research*, ed. D.A. Otto. Washington D.C.: U.S. Government Printing Office, pp. 322–344.

Kihlstrom, J.F. (1996). Perception without awareness of what is perceived, learning without awareness of what is learned. In *The Science of Consciousness: Psychological, Neuropsychological, and Clinical Reviews*, ed. M. Velmans. London: Routledge, pp. 23–46.

Kornhuber, H.H. and Deecke, L. (1965). Hirnpotentialänderungen bei Willkürbewegungen und passiven Bewegungen des Menschen: Bereitschaftspotential und reafferente Potentiale. *Pflügers Archiv für die gesamte Physiologie des Menschen und der Tiere* 284: 1–17.

Lenneberg, E.H. (1967). *Biological Foundations of Language.* New York: Wiley.

Libet, B. (1985). Unconscious cerebral initiative and the role of conscious will in voluntary action. *Behavioral and Brain Sciences* 8: 529–566.

Libet, B. (1996). Neural processes in the production of conscious experience. In *The Science of Consciousness: Psychological, Neuropsychological, and Clinical Reviews*, ed. M. Velmans. London: Routledge, pp. 96–117.

Libet, B., Wright Jr., E.W., Feinstein, B. and Pearl, D.K. (1979). Subjective referral of the timing for a conscious experience: A functional role for the somatosensory specific projection system in man. *Brain* 102: 193–224.

Mangan, B. (1993). Taking phenomenology seriously: The "fringe" and its implications for cognitive research. *Consciousness and Cognition* 2(2): 89–108.

McMahon, C.E. & Sheikh, A. (1989). Psychosomatic illness: A new look. In *Eastern and Western Approaches to Healing*, ed. A. Sheikh & K. Sheikh. New York: Wiley-Interscience, pp. 296–324.

Pelletier, K.R. (1993). Between mind and body: Stress, emotions, and health. In *Mind Body Medicine: How to Use Your Mind for Better Health*, ed. D. Goleman & J. Gurin. New York: Consumer Reports Books, pp. 19–38.

Searle, J. (1994). Intentionality (1). In *A Companion to the Philosophy of Mind*, ed. S. Guttenplan. Oxford: Blackwell, pp. 379–386.

Sheikh, A.A., ed. (2001). *Healing Images: The Role of Imagination in the Healing Process.* Amityville, New York: Baywood Publishing Company.

Sheikh, A.A., Kunzendorf, R.G. & Sheikh, K.S. (1996). Somatic consequences of consciousness. In *The Science of Consciousness: Psychological, Neuropsychological, and Clinical Reviews*, ed. M. Velmans. London: Routledge, pp. 140–161.

Skrabanek, P. & McCormick, J. (1989). *Follies and Fallacies in Medicine.* Glasgow: The Tarragon Press.

Stoffregen, T.A. & Benoît, G.B. (2001). On specification of the senses. *Behavioral and Brain Sciences* 24(2): 195–261.

Syrjala, K.A. & Abrams, J.R. (1996). Hypnosis and imagery in the treatment of pain. In *Psychological Approaches to Pain Management: A Practitioner's Handbook,* ed. R.J. Catchel & D.C. Turk. New York: The Guildford Press, pp. 231–258.

Tye, M. (1995). *Ten Problems of Consciousness: A Representational Theory of the Phenomenal Mind.* Cambridge, Mass: MIT Press.

Velmans, M. (1990). Consciousness, brain, and the physical world. *Philosophical Psychology* 3: 77–99.

Velmans, M. (1991a). Is human information processing conscious? *Behavioral and Brain Sciences* 14(4): 651–669.

Velmans, M. (1991b). Consciousness from a first-person perspective. *Behavioral and Brain Sciences* 14(4): 702–726.

Velmans, M. (1993). Consciousness, causality and complementarity. *Behavioral and Brain Sciences* 16(2): 409–416.

Velmans, M., ed. (1996a). *The Science of Consciousness: Psychological, Neuropsychological and Clinical Reviews.* London: Routledge.

Velmans, M. (1996b). Consciousness and the "causal paradox". *Behavioral and Brain Sciences* 19(3): 537–542.

Velmans, M. (1998). Goodbye to reductionism. In *Towards a Science of Consciousness II: The Second Tucson Discussions and Debates*, ed. S. Hameroff, A. Kaszniak & A. Scott. Cambridge, Mass: MIT Press, pp. 45–52.

Velmans, M. (2000). *Understanding Consciousness*. London: Routledge/Psychology Press.

Velmans, M. (2001a). A natural account of phenomenal consciousness. *Consciousness and Communication* 34(1–2): 39–59.

Velmans, M. (2001b). Heterophenomenology versus Critical Phenomenology: A Dialogue with Dan Dennett. http://cogprints.soton.ac.uk/documents/disk0/00/00/17/95/index.html.

Velmans, M. (2002a). How could conscious experiences affect brains? *Journal of Consciousness Studies* 9(11): 3–29.

Velmans, M. (2002b). Making sense of the causal interactions between consciousness and brain. *Journal of Consciousness Studies* 9(11): 69–95.

Velmans, M. (2003). *How Could Conscious Experiences Affect Brains?* Exeter: Imprint Academic.

Wall, P.D. (1996). The placebo effect. In *The Science of Consciousness: Psychological, Neuropsychological and Clinical Reviews,* ed. M. Velmans. London: Routledge, 162–180.

Watkins, A. (1997). Mind-body pathways. In *Mind-Body Medicine: A Clinician's Guide to Psychoneuroimmunology*, ed. A. Watkins. New York: Churchill Livingstone, pp. 1–25.

Watt, D. (2000). The centrencephalon and thalamocortical integration: Neglected contributions of periaqueductal gray. *Consciousness and Emotion* 1(1): 91–114.

V. CONSCIOUSNESS, INTERPRETATION, AND HIGHER-ORDER-THOUGHT

DAVID M. ROSENTHAL

1. THE PROBLEM OF EXPLAINING CONSCIOUSNESS

Few contemporary researchers in psychology, philosophy, and the cognitive sciences have any doubt about whether mental phenomena occur without being conscious. There is extensive and convincing clinical and experimental evidence for the existence of thoughts, desires, and related mental states that aren't conscious.

We characterize thoughts, desires, intentions, expectations, hopes, and many other mental states in terms of the things they are about and, more fully, in terms of their content, as captured by a sentence nominalization, such as a clause beginning with the word 'that'. The philosophical literature follows Franz Brentano's adaptation of Thomist terminology in referring to all such states as intentional states.

But there is another type of mental phenomena, which lack intentionality and whose mental nature consists instead of some qualitative feature. These states include bodily sensations, such as aches and pains, and perceptual states, such as visual sensations of color and tactile sensations of heat and cold. And these states all exhibit some mental quality or another, such as the mental quality distinctive of pain or the mental quality of red or blue.[1] And even theorists who acknowledge that intentional states can and do occur without being conscious have sometimes insisted that qualitative states cannot. There is, according to these theorists, nothing to a state's being qualitative or exhibiting some mental quality unless that state is conscious – unless it is, as we might metaphorically say, "lighted up".

It's striking that Freud himself seems to have adopted this double standard toward the two types of mental state. In his metapsychological paper, "The Unconscious", for example, he writes that "all the categories which we employ to describe conscious mental acts, such as ideas, purposes, resolutions, and so forth, can be applied to [unconscious mental occurrences]" (Freud 1915e, p. 168). But he seems here to have in

[1] It is crucial to distinguish the mental qualities characteristic of perceptual sensations from the distinct, but corresponding physical qualities characteristic of the objects perceived. On this, see Rosenthal (1999a).

mind only purely intentional states, which altogether lack qualitative character.

Later in that same paper, by contrast, he writes that "[i]t is surely of the essence of an emotion, that we should be aware of it. [...] Thus the possibility of [...] unconsciousness would be completely excluded so far as emotions, feelings and affects are concerned" (Freud 1915e, p. 177; see also 1923b, pp. 22–3; 1940a, p. 197). To describe feelings as unconscious, he concluded, is to "speak in a condensed and not entirely correct manner" (Freud 1923b, p. 22) about cases in which the representational character of the feelings is repressed or misrepresented (Freud 1915e, pp. 177–8).[2]

But such a double standard is unjustified. There is ample evidence that qualitative states, no less than intentional states, occur without being conscious. In masked-priming experiments, subjects presented very briefly with two successive stimuli report being wholly unaware of the first stimulus, even though it has a demonstrable effect on mental processing. And blindsight subjects, part of whose primary visual cortex has been destroyed, report that they don't see visual stimuli in the relevant area of the visual field, but their forced-choice guesses about visible characteristics of such stimuli are startlingly accurate. Because these effects correspond to differences among mental qualities in the absence of conscious sensing, they point to the occurrence of qualitative states that aren't conscious.[3]

Why, then, do many theorists continue to deny the occurrence of nonconscious qualitative states? One reason stems from a conflation of two ways we use the term 'conscious'. We use that term to mark a difference between two conditions that a person or other creature might be in and also to mark a difference between two ways that mental phenomena occur.

Consider first mental phenomena. When somebody has some thought, desire, perception, or feeling but is in no way whatever aware of that state, the state plainly is not a conscious state.[4] It follows that conscious states are those we are conscious, in some suitable way, of

[2] Freud did allow, however, that something corresponding to conscious qualitative states can occur without being conscious (Freud 1923b, p. 22). So it may be that the issue here is simply verbal, hinging not what states can occur but on how one uses the terms 'feeling' and 'sensation'.

[3] On masked priming see Marcel (1983a, 1983b). On blindsight, see Weiskrantz (1997, 1986) and Merikle et al. (2001).

[4] I use 'aware of' and 'conscious of' interchangeably here.

being in. In section II I'll take up the question of just what that suitable way might be.

There is no circularity in explaining what it is for a mental state to be conscious by reference to one's being conscious *of* that state. We understand what it is to be conscious *of* something independently of any appeal to mental states' being conscious or not. We're conscious of something when we see it or hear it or sense it in some other way, and we're conscious of things when we have thoughts about them as being present.

In addition to using the term 'conscious' to mark the difference between those mental states we are suitably aware of and those we are not, we also use that term to mark a difference in two conditions that people and other creatures may be in. A person or other animal is conscious if it's awake and mentally responsive to sensory stimulation. This is plainly a different property from that of a mental state's being conscious. I'll use the terms *creature consciousness* and *state consciousness* to refer, respectively, to the property a creature has when it's conscious and the property a mental state sometimes has of being conscious.

It's obvious that creature consciousness does not imply state consciousness, since people while awake are typically in many mental states that aren't conscious. But creature consciousness involves a creature's being mentally responsive to sensory stimulation. So it may be tempting to think that creature consciousness implies at least the consciousness of whatever sensory states a creature is in when that creature is conscious. And perhaps, then, all qualitative states whatever are conscious. A little reflection, however, will dispel that illusion. By itself, a creature's being conscious cannot preclude its sensing some things nonconsciously. Indeed, a creature might be awake and mentally responsive to sensory stimulation without any of its qualitative states being conscious. This never happens in the normal human case, but it might well occur with other creatures, fewer of whose mental states are conscious.

A second explanation for the widespread idea that qualitative states cannot occur without being conscious stems from a feature of perceptual sensations that's unique among mental states. Intentional states, such as thinking, desiring, anticipating, doubting, wondering, and the like can all be expressed in speech, as well as in various forms of nonverbal behavior. The same holds for affective states, such as anger, joy, sadness, and fear, which themselves are in part intentional; when we are angry, joyous, sad, or fearful, we are angry, joyous, sad, or fearful that something is the case, and we can express these feelings both verbally and nonverbally.[5] Even bodily sensations, such as pains, tickles, and itches, are expressible, by saying 'ouch' or writhing, by laughing, and by scratching.

By content, we cannot express our perceptual sensations, such as visual sensations of red or auditory sensations of a violin, either in speech or nonverbally. We express our perceptions of these things, which have intentional content, for example, by saying that something is red or that it's a violin. But, since perceptual sensations have only mental qualities, and no intentional content, there is accordingly no way to express them. And, because we cannot express these states, it's inviting to think that the only way we can detect their presence is by being conscious of them. The purely qualitative character of perceptual sensations, it may seem, means that their only detectable effect is in being conscious. And that makes it seem, in turn, as though there is nothing to a perceptual sensation's occurring unless it's conscious. And it may then seem natural to extrapolate to all states that have qualitative character of which sort, including bodily sensations and emotions.

But the absence of any way to express perceptual sensations does not actually imply any such result. Though the purely perceptual qualitative character of perceptual sensations prevents them from having, by themselves, any effect on behavior, they do have effects on other mental state states and processes. They characteristically lead, for example, to our having full-fledged perceptions with intentional content and to our having perceptual beliefs. Perceptual sensations are detectable, even when they aren't conscious, by their causing other mental states. Indeed, this is exactly how the occurrence of nonconscious perceptual sensations in blindsight and masked priming is experimentally detected. The inexpressibility of perceptual sensations does not support the idea that qualitative states cannot occur without being conscious.

There is a third, somewhat more theoretical reason why many writers have held that qualitative states cannot occur without being conscious, which stems from a familiar idea about how to reconcile our commonsense picture of reality with that of modern physics. Physics requires that the properties of physical objects can all be described in mathematical terms, but it seems that commonsense physical qualities, such as color and sound, resist such mathematical treatment. A standard solution is simply to deny that color and sound, conceived as common

[5] I return to the expressibility in speech and nonverbal behavior of cognitive and affective states in section 4.

sense does, are properties of physical objects at all, and insist that these qualities occur instead only as mental properties.

This manoeuvre has a striking, if largely unnoticed, side effect. Commonsense physical qualities seem problematic only when we're conscious of them. So we will be tempted to relocate as mental properties only those physical qualities that we're conscious of. But that means that the relocated mental qualities will have consciousness already built in. This device for reconciling common sense with physics actually distorts our very conception of mental reality.[6]

Adjusting the way we square common sense with mathematical physics avoids this effect. We can, instead, simply identify the physical qualities of common sense with physical properties that a mathematical physics can accommodate, colors, for example, with suitable surface reflectance properties.[7] We can then explain the commonsense character of such properties by appeal to the way we're conscious of them, without thereby supposing that this commonsense character is part of the properties we're conscious of. And, once we stop construing mental qualities as relocated versions of physical qualities, there is no remaining reason to see consciousness as automatically built into all mental qualities.

2. HOW WE ARE CONSCIOUS OF OUR CONSCIOUS STATES

A mental state is conscious when the individual that's in that state is, in some suitable way, conscious of being in that state. So the main job of a satisfactory theory of what I've called state consciousness is to say just what that suitable way is.

There are two traditional answers to this question. One is the view, often associated with Brentano, that the consciousness of our conscious states is intrinsic to the states themselves. Every mental state, according to Brentano, is not only about its ostensible object; it is also about itself. As Brentano he put it, all mental acts "apprehend [themselves,] albeit indirectly" (Brentano 1884, p. 128). It is inviting to wonder whether the view that all intentional states have dual content, which Freud would certainly have heard in Brentano's weekly lectures, may have influenced Freud's own distinction between the manifest and latent content of thoughts and desires.

---

[6] For more on this see Rosenthal (1999b).

[7] More precisely, suitable ratios of such properties.

The other traditional view is that we're conscious of our conscious states by way of a kind of "inner sense", in Kant's useful phrase;[8] we're conscious of our conscious states by sensing them (Kant 1787, A22, B37, p. 174). It's likely that Freud held some version of this view; "consciousness", he wrote, plays the role "of a sense-organ for the perception of mental qualities" (Freud 1900a, p. 615).[9] Both these traditional models, however, have important failings, though these difficulties combine to point the way toward a more successful theory. Consider first the inner-sense model. One reason it seems inviting to hold that qualitative states are conscious when we sense them is the idea that such higher-order sensing might result in the conscious qualitative states' having their "lighted up" qualitative character. The higher-order sensing would itself contribute the relevant conscious quality. But even if that were so, this wouldn't apply to our conscious thoughts, desires, wishes, hopes, and other intentional states; it's highly implausible that we sense those states in any way whatever.

But there is a deeper and more decisive problem for the inner-sense model. Sensing consists in being in states that have some qualitative character or other. That's why higher-order sensing seems to help with conscious qualitative states but not conscious intentional states. But, when a qualitative state is conscious, only one qualitative character seems to occur, that of the state that's conscious. There doesn't, in addition, seem to be a second qualitative property that belongs to some higher-order sensation. But qualitative character is distinctive of sensations. So, if there's no higher-order qualitative character, there's also no higher-order sensation. As inviting as the inner-sense model may initially seem, it cannot be sustained.

There is another, related difficulty. Every sensation not only has some distinguishing qualitative character, but it also belongs to a particular sensory modality, such as vision, hearing, and so forth. But, as

---

[8] John Locke uses the term "internal Sense" (1700, 2, i, 4, p. 105); cp. his remark that "[c]onsciousness is the perception of what passes in a Man's own mind" (1700, 2, i, 19, p. 115). Locke also seems sometimes to holds that our consciousness of our conscious states is an intrinsic aspect of those states, as when he writes that "thinking consists in being conscious that one thinks" (1700, 2, i, 19); so perhaps he held that we sense our conscious states by way of an intrinsic aspect of those states. See also Armstrong (1980, p. 61) and Lycan (1996).

[9] I have used 'mental' in place of Strachey's somewhat anachronistic 'psychical'. On Freud's perceptual model of consciousness, see also Solms (1997a, 1997b) and Rosenthal (1997a).

Brentano noted, following Aristotle, if we sense our conscious qualitative states, it's not obvious what modality the higher-order sensation could have. The modality of the first-order state won't do, since that modality senses physical qualities, not sensations, and no other modality seems available (Brentano 1884, p. 127).[10]

Brentano actually invokes this very consideration in support of his alternative model of consciousness. If we're not aware of our conscious states by distinct, higher-order sensations of those states, our awareness of our conscious states must, he thought, be due instead to something intrinsic to the states themselves. Whether one's awareness of a mental state counts as being intrinsic to that state will of course hinge on how we individuate mental states generally. And it's not immediately obvious what standard of individuation we could use that wouldn't beg the question at hand.

I'll return shortly to the question of individuating mental states. But I want now to note a more serious difficulty that faces Brentano's intrinsicalist model. Many mental states occur without being conscious. Brentano recognized that a mental state's occurring without being conscious is conceptually coherent (Brentano 1884, p. 128),[11] but he denied that this ever happens. But we know that it does; so we must explain how it is that some mental states occur consciously and others not. Building consciousness into the state itself makes that difficult, if not impossible. Mental states sometimes shift between being conscious and not being conscious; how can the intrinsicalist story explain that? Intrinsicalism goes naturally with the outmoded view that mental states never occur without being conscious.[12]

It's clear that Freud rejected any such intrinsicalist view. If a state's being conscious does consist in one's being conscious of it, it's natural to expect that it could occur without one's being conscious of it and,

---

[10] Aristotle's argument, which Brentano adapts, assumes that we do somehow sense that we see. If so, and if the sense we use to sense that we see is distinct from the sense of sight, then that other sense and the sense of sight would both have color as their proper object. But distinct senses never share the same proper object (Aristotle, de Anima III 2, 425b12-4).

[11] "An unconscious consciousness is no more a contradiction in terms than an unseen case of seeing" (Brentano 1884, p. 102), and "[a] presentation of a sound without a presentation of the act of hearing would not be inconceivable" (Brentano 1884, p. 128).

[12] Indeed, Brentano's intrinsicalist view seems to be is his only reason for holding that mental states are always conscious.

hence, without its being conscious. The default condition for mental states will thus be for them to occur nonconsciously. Freud explicitly drew this conclusion, writing that "[t]he mental, whatever its nature may be, is in itself unconscious." Consciousness is "an inconstant quality" of mental occurrences, "one that is far oftener absent than present" (Freud 1940b, p. 238).

Since the default condition for mental states is that they occur nonconsciously, no mechanism of repression is needed for such nonconscious occurrence of mental states. In and of itself, every mental state occurs without being conscious. A repressed state is one that occurs consciously, because an accompanying higher-order state occurs in virtue of which one is conscious of the state, but the mechanism of repression then blocks that state's being conscious. So that mechanism must interfere with the higher-order state in virtue of which the repressed state would otherwise be conscious.

Let's take stock. The inner-sense model fails because there are no higher-order qualities, distinct from the first-order qualities of the states we're conscious of, and we know of no modality that such higher-order qualities could belong to. The intrinsicalist picture fails because we have no independent way to individuate mental states that yields the right result and, more important, if consciousness is intrinsic to mental states, we cannot explain why mental states are sometimes conscious and sometimes not.

Mental states are conscious when we're conscious of them. There are only two ways we know of in which we're conscious of things, by sensing them and by having thoughts about them as being present. Since we aren't conscious of our conscious states by sensing them or by way of something intrinsic to the state, the only remaining possibility is that we're conscious of them by having thoughts about them as being present, thoughts that are extrinsic to and distinct from the conscious states, themselves.

We can say more about what these higher-order thoughts (*HOTs*) must be like. When a mental state is conscious, one is conscious of oneself as being in that state. So the content of a HOT must be that one is in that very state.[13] HOTs need not conceptualize their target states in any very elaborate way; a mental state can be conscious without one's being

[13] More precisely, it must be that I, myself, am in that state; it's not enough for me simply to ascribe the state to an individual that happens, unbeknownst to me, to be me. On this and related issues, see Rosenthal (2003).

conscious of it in the sophisticated way adult humans think about mental states. Thus children may well lack adult human concepts for thoughts, sensations, feelings, and even the mental itself.[14] All that's needed is that the HOT make one aware of the state in respect of some suitable property, a property that we adult humans would classify as mental.

We are seldom aware of any such HOTs. But that's no problem for the model. We wouldn't be aware of any HOT unless it was conscious, and that will happen only if a HOT is in turn accompanied by another HOT about it, in effect, a third-order thought. This is rare, but it does sometimes happen. There is a difference between mental states' being conscious in the ordinary, everyday way and their being introspectively or reflectively conscious. And mental states are reflectively conscious when they are accompanied not simply by HOTs, but by HOTs that are themselves conscious. The reflective character of introspective consciousness results from one's actually being conscious of being aware that one is in some particular state. Since HOTs are typically not conscious, moreover, we avoid the regress that would occur if we explained one state's being conscious by appeal to another that must itself be conscious.

When a mental state is conscious, one is conscious of that state in a somewhat special way that many theorists have characterized as direct or immediate. Thus Descartes, who builds consciousness into his concept of a mental state, insists that "the word 'thought' applies to all that exists in us in such a way that we are immediately conscious of it" (Descartes 1641, p. 160).

This apparent immediacy doubtless explains much of the intuitive appeal of the intrinsicalist and inner-sense models. One's awareness of a mental state would plainly be immediate if it were intrinsic to that state, and sensing things also seems, intuitively, to be direct and unmediated. But the HOT model can also explain this sense of immediacy. We can require that HOTs rely on no conscious inference, no inference, that is, of which one is conscious. Suppose I come to have a thought that I feel a certain way solely because you tell me and I have confidence in you; that HOT won't make the feeling conscious. For the feeling to be conscious, I must have a HOT that's independent of any such conscious inference. It's this independence of conscious inference that results in

[14] For developmental work that suggests this, see (Perner 1991; Perner, Leekam & Wimmer 1987; Wellman 1990; Wimmer & Perner 1983). For an alternative view see Fodor (1992), and a reply by Perner (1995).

the subjective sense that nothing mediates between the feeling and my awareness of it.

This raises a theoretically delicate issue. Suppose you tell me that I feel a certain thing and, though it doesn't subjectively seem to me as though I do, I believe you. I come to have a HOT based on consciously inferring from what you said. My feeling isn't conscious, just as the theory provides. But suppose now that my feeling does become conscious; this presumably often happens in psychoanalytic treatment. How can the HOT model explain this shift?

Originally my HOT was based on a conscious inference; I had the thought that I had a certain feeling, and I also consciously thought I had that feeling because you'd said I had it. Later, even though I may consciously remember what you'd said, it comes to seem to me that I'm conscious of the feeling independently of your having said anything. My thought that I have that feeling comes to seem, subjectively, to be independent of your remark. Since my HOT is now independent of any conscious inference, the feeling now is conscious. In what follows, I'll assume these qualifications, using the term 'HOT' to apply only to the noninferential cases and the term 'noninferential' to mean independence of inference of which one is conscious. And I'll return to these issues in closing.

HOTs that rely on conscious inference and those which do not are not two kinds of HOT; the very same HOT might at one point rely on conscious inference and at another point not. The difference is wholly extrinsic to the HOT, itself. Nor, therefore, is the difference a matter of the genesis of the HOT, since a HOT might first occur as a result of conscious inference and subsequently come to be independent of the conscious inference. All that matters is whether it seems, subjectively, as though one would be conscious of the target state even in the absence of any such inference. It does not even matter whether one actually would be conscious of the state even without any such inference, but only whether it seems that way.[15]

[15] On so-called conceptual-role theories of intentional content, a state's content is a matter of its inferential connections with other states. So one might object that the HOT model conflicts with a conceptual-role account of content, since HOTs would vary in content depending of whether they are based on conscious inference. (I am indebted to Pierre Jacob, personal communication, for this ingenious objection).

But conceptual-role theories characterize content not as a function not of the actual inferences one makes but as a function of dispositions to infer, of what inferences

HOTs cannot be dispositional,[16] since being disposed to have a thought about something doesn't make one conscious of that thing. More important, HOTs cannot be internal to the states they are about.[17] So the HOT model cannot be combined with Brentano's intrinsicalist view. The reason derives, ironically, from something that Brentano noted, namely, that we individuate intentional states by reference to the mental act being performed (Brentano 1884, p.127). But we individuate mental acts, in turn, partly by reference to the type of mental attitude involved, such as believing, desiring, doubting, wondering, expecting, and the like.

Suppose, then, that the HOT in virtue of which a mental state is conscious is internal to that state, and suppose also that the state in question is a case of wondering something. Wondering about something doesn't make one conscious of that thing; an intentional state makes us conscious of the things it is about only when that state has an assertoric mental attitude. So the HOT in virtue of which our state of wondering is conscious must itself be assertoric. But, since the HOT will then have a different type of mental attitude from its target, the HOT and its target will be distinct states. HOTs cannot be internal the states they are about.[18]

It's worth distinguishing this view from the position put forth independently by Daniel C. Dennett and Bernard Baars that a mental state's being conscious consists in its being widely broadcast within our cognitive system. Thus Dennett writes:

"Consciousness is cerebral celebrity. [...] Those contents are conscious that persevere, that monopolize resources long enough to achieve certain typical and "symptomatic" effects – on memory, on the control

---

one might make. And HOTs will bear the same inferential relations so construed. Moreover, what matters to the HOT model is only whether the actual inferences are conscious, and that wouldn't in any case matter to a conceptual-role account of content.
For conceptual-role theories, see Harman (1973, 1999, part III) and Block (1986).

[16] As Peter Carruthers has argued they are; see, e.g., Carruthers (2000).

[17] As argued by Gennaro (1995) and, more recently, Kriegel (2002, 2003).

[18] This may recall the difficulty the inner-sense model faced about a possible mismatch of sensory modality between the higherorder and the target sensings. For more on the HOT hypothesis, see Rosenthal (1986, 1997b, 2002).

of behavior and so forth" (Dennett 1993, p. 929; see also Dennett 1991; Baars 1998, 1997).[19]

Doubtless many conscious states satisfy this description. But many states that meet the description are nonetheless not conscious. A state can be widely broadcast to cerebral areas, monopolize resources, and have the characteristic effects Dennett mentions without being conscious. Indeed, exactly that presumably occurs with many states that guide and influence our everyday behavior. Such states widely influence both cerebral functioning and behavior without our being in any way whatever conscious of them. Indeed, it is tempting to think that cerebral broadcasting corresponds more to Freud's notion of the preconscious than to any intuitive notion of a mental state's being actually conscious. Similarly, a state can be conscious despite its having little such influence. Though they often go together, consciousness is not cerebral celebrity.

As noted earlier, Freud evidently adopted a version of the sensory model of what it is for mental states to be conscious states. It's not clear, however, how that view squares with his conviction that qualitative states are always conscious. It seems to result in the kind of regress just mentioned. The states in virtue of which we perceive things all exhibit some mental quality. On the perceptual model, a mental state's being conscious consists in its being accompanied by a higher-order qualitative state. But, if qualitative states never occur without being conscious, that higher-order qualitative state will in turn have to be accompanied by a still higher-order qualitative state, and so forth without limit. Per-haps it was this kind of consideration that led Freud to write that "the fact of consciousness" "defies all explanation or description" (Freud 1940b, p. 282).

3. CONSCIOUSNESS AND CONFABULATION

Theorists of a traditional Cartesian bent hold that mental states are all conscious. But often they also insist on two other views, which go beyond that Cartesian dictum, namely, that consciousness represents our mental states infallibly and that it represents those states exhaustively.

[19] Ned Block's notion of access consciousness is also of this sort. A state is access conscious if it is "poised to be used as a premise in reasoning, [...] [and] for [the] *rational* control of action and [...] speech" (Block 1995, p. 231).

These three views all stem from a fundamental conviction that the mind is transparent to itself, in Gilbert Ryle's apt, disparaging metaphor, that the mind is "self-intimating" (Ryle 1949, p. 158).[20] And the mind's being transparent to itself would ensure not only that all states of the mind are conscious, but also that consciousness invariably represents those states accurately and fully.

The HOT model denies all these things. Not only are mental states not all conscious; we have, in addition, no reason to suppose that HOTs will represent the mental states they are about in respect of all their mental properties, nor even that they will always represent those mental targets accurately.

This is all to the good. Once the view that mental states are all conscious loses its grip on us, the views about exhaustiveness and infallibility should strike us, in the words of C. D. Broad, as "a curious superstition" (Broad 1925, p. 284).[21] And there is ample evidence that neither view is correct.

Consider first exhaustiveness. Suppose I consciously see something red. In most ordinary circumstances my sensation won't be conscious in respect of that specific shade, but only in respect of some generic red. But attending to the sensation often results in the sensation's becoming conscious in respect of some far more specific shade. This gives us reason to think that, even when the sensation is conscious only as being generically red, the mental character of the sensation itself is typically far more specific. The mental quality of the sensation is all along the more specific shade, of which we're conscious sometimes only generically and at other times with greater specificity. Consciousness does not invariably represent our mental states fully and in respect of all their specific mental properties; indeed, it presumably does so relatively seldom.

How about infallibility? Even if consciousness fails to represent everything mental about our conscious states, perhaps it is, still, reliable about the mental properties it does represent our conscious states as having. But there is reason to reject this as well. Clinical contexts provide examples of consciousness misrepresenting which mental states

[20] And, for the case of qualitative states, that the mind is "self-luminous" (Ryle 1949, p. 159).

[21] Broad's remark was specifically about the information that introspective consciousness delivers, and not about ordinary, unreflective consciousness, but the same point applies.

one is in, and there is also experimental evidence from social psychology. Subjects sometimes report having beliefs and desires when there is independent evidence that they don't actually have them, especially when those beliefs and desires fit with an explanatory or motivational story the subjects want to adopt (Nisbett & DeCamp Wilson 1977; Greenwald & Banaji 1995; White 1988; Wilson, Hodges & LaFleur 1995).

Such confabulation can occur not only with intentional states, such as beliefs and desires, but even with qualitative states. A dramatic example involves dental patients, who sometimes report pain when drilling occurs in a tooth in which pain cannot occur, because the nerve is anesthetized or dead. The standard explanation, known as dental fear, is that the patient has sensations of vibration and fear, and misconstrues the unpleasant experience as pain. Giving the patient this explanation typically results in the patient's no longer reporting pain when drilling resumes, but rather sensations of vibration and fear. Still, the patient also continues to have a vivid memory of the earlier conscious experience as being pain. Consciousness sometimes actually misrepresents our conscious qualitative states.[22]

The HOT hypothesis not only accommodates these phenomena; it predicts them. The reason a mental state is conscious when it's accompanied by a HOT is that one's HOT, in representing one as being in that state, makes one conscious of being in it. In having a HOT about a state, one is conscious of oneself as being in that state. But, as a general matter, having a thought makes one conscious of something only in respect of those properties which the thought attributes to that thing. When I have a thought about some object as being present, I'm conscious of that object in just the way that thought describes it. Being conscious of it in any other way would require my having a thought or sensation that represented the object in that other way.

HOTs are simply a special case of this. One is conscious of a conscious state in respect of just those properties which the HOT represents that state as having. When one has a conscious sensation of red, for example, the way one is conscious of the mental qualities of that sensation is a matter of how one's HOT represents it. So the very same sensation of red can be conscious in respect of a relatively generic redness or a more specific shade, depending on how one's HOT describes that

[22] For evidence of related phenomena, see Staats, Hekmat & Staats (1998) and Holmes & Frost (1976).

sensation, depending, that is, on what mental quality the HOT attributes to the sensation. And, since no thought captures all the properties of the things it's about, no HOT will represent all the mental properties of its target mental state. Consciousness is not exhaustive.

Similarly, the HOT model also predicts that consciousness is not infallible. Not all our thoughts represent things accurately, and those which don't can make us conscious of things in inaccurate ways. So it is with HOTs. In having an inaccurate HOT, one is conscious of oneself as being in a particular state under an inaccurate description. This kind of thing presumably happens in clinical situations. One may, for example, be motivated not to be conscious of oneself as having a certain desire or feeling. And one's motivation not to be conscious of oneself as having that desire or feeling may lead one to misrepresent the state and to be conscious of oneself as having, instead, a somewhat different desire or feeling. On the HOT model, one's HOTs misdescribe the relevant state so as to make it less unacceptable. One is conscious of the desire or feeling as being a state of a different sort from that which one wishes to avoid acknowledging. It is because this sort of thing can happen that it's open for Freud to claim that the feelings we call unconscious do not literally fail to be conscious. Rather one is conscious of those states in respect of their qualitative character of the feeling but not their intentional content.[23]

In the extreme case, one's HOT might even represent one as being in a mental state that one isn't in at all. But it's important to note that this cases differs only in degree from the case just considered, in which a HOT represents one as being in a state but inaccurately characterizes the state. Strictly speaking, when a thought is inaccurate, the thing that thought purports to be about doesn't actually exist.

What matters for conscious subjectivity, then, is how one's HOTs represent things, that is, what states they represent one as being in. It doesn't matter for conscious subjectivity whether one is actually in that state. Consider two cases. In the first, one is in a particular mental state and one also has an accompanying HOT that correctly characterizes that state; in the second, one has that very same HOT but the state is absent. In both cases one is conscious of oneself as being in that state. So the two cases will be subjectively indistinguishable. There will doubtless be functional differences between the two cases; every mental state has

[23] Or, in any event, not the intentional content they actually have. See Freud (1915e, pp. 177-8; 1923b, pp. 22-3; 1940a, p. 197).

causal connections with behavior and with other mental states. So, if the target state is absent, that will be detectable. Still, there will be no difference between the two cases so far as conscious subjectivity is concerned.

4. CONSCIOUSNESS, SPEECH, AND INTERPRETATION

A major advantage of the HOT model is that it explains how mental states can be conscious in respect of one of its mental properties but not another. A mental state is never simply conscious, *tout court*; it is conscious *as* a state of a particular mental type. The model explains this because a HOT specifies the mental type in respect of which one is conscious of a state.

HOTs, moreover, do a far better job than inner sense could in capturing the way mental states are conscious in respect of various mental properties. Sensing represents things only by way of properties that are specific to the several sensory modalities, whereas there are no limits to the ways thoughts can represent things. So thoughts are far more flexible than sensing would be in capturing variations in the way mental states are conscious.

Another major advantage of the HOT model is that it explains the close and important connection that holds in humans between consciousness and speech. For creatures, such as adult humans, who can readily say what mental states they are in, a state's being conscious coincides with one's being able to report noninferentially that one is in that state. This connection is widely accepted in experimental psychology as a mark in humans of a state's being conscious.[24]

I report a mental state when I say that I am in that state. But every sincere, meaningful speech act expresses a thought with the same content as the speech act and a mental attitude that parallels its force. That's why one cannot meaningfully say things such as 'It's raining but I don't think it is'; the first part purports to express a thought that the second part denies the existence of.[25] So a noninferential report that one is in a

[24] See, e.g., Marcel (1983a, 1983b). Writers seldom if ever make explicit that the report must be noninferential, but it's plainly understood; nobody would hold that reportability that relied on conscious inference coincides with consciousness.

[25] This kind of construction was first noted by G. E. Moore, who pointed out that, although it is not contradictory, one cannot make any statement with such a

mental state always expresses one's noninferential thought that one is in that state. It expresses, that is, a HOT.

It follows that the ability to report noninferentially that one is in a mental state is the same as the ability to express verbally a noninferential HOT to the effect that one is in that state. This allows us to explain why we can noninferentially report any conscious mental state. We can do so because, when a mental state is conscious, it's accompanied by the very HOT that a report of that state would express. Similarly, the reason we cannot report any mental state that isn't conscious is that the HOT that such a report would express doesn't occur. The HOT hypothesis is the best explanation of the connection between consciousness and noninferential reportability.[26]

The HOT hypothesis explains as well another striking aspect of the connection between speech and consciousness, namely, that whenever we verbally express a thought, that thought is conscious. When I say that it's raining, I verbally express my thought that it's raining. I could also express that very same thought nonverbally, for example, by taking my umbrella as I go out.[27] When I express my thought nonverbally, the thought may well remain unconscious; I may take my umbrella without thinking consciously about it. But, when I express the thought verbally, by saying that it's raining, the thought is always conscious. Why is that? Simply positing a necessary connection between consciousness and speech, as some traditional thinkers have (see, e.g., Descartes, Letters to Newcastle and to More, 1645 & 1649), amounts to giving no explanation at all.

Whenever I say it's raining, I could as easily have said that I think it's raining. The two statements do not, of course, mean the same; one is about the weather and the other is about my mind. But they do have virtually the same conditions of use.[28] Even more important for our pur-

---

construction (Moore 1942, p. 543; 1944, p. 204). Following Wittgenstein (1953, 2, §10), this type of construction is known as Moore's paradox. See also Rosenthal (1995).

[26] For more on this, see Rosenthal (1993).

[27] One might urge that nonverbal behavior never expresses the thoughts that underlie such behavior in the same way that speech acts do; I am grateful to Josef Perner for pressing this point, in conversation. But that issue does not affect the present argument.

[28] Perhaps saying that I think it's raining conveys a measure of caution not present when I simply say it's raining, but that difference doesn't affect the present argument.

poses, that equivalence in conditions of use is automatic and second nature for us, so much so that we may not even notice on any given occasion which of the two we said, nor which of the two another person said.[29]

Suppose, then, that I say that it's raining. Because I might as easily have said instead that I think it's raining, I must have had the thought that this other remark would have expressed; otherwise the interchangeability of the two statements wouldn't be fully automatic and second nature. But the thought I would express by saying that I think it's raining would be the HOT that I think it's raining. And, since the automatic interchangeability ensures that I have that HOT, my first-order thought that it's raining is itself a conscious thought . The HOT hypothesis explains why, whenever I say anything at all, the thought I thereby express is conscious.[30]

Things are different when one expresses a thought nonverbally. There is no equivalence in performance conditions between the nonverbal expression of a thought and any expression, verbal or otherwise, of the corresponding HOT. So the nonverbal expressing of a thought gives us no reason to posit an accompanying HOT. The HOT model thus also explains why nonverbally expressed thoughts need not be conscious.[31]

One might object that parapraxis shows that verbally expressed thoughts are not, after all, invariably conscious. Suppose that somebody, intending consciously to declare a meeting open, slips and says 'I declare this meeting closed'.[32] One could interpret this speech performance

---

[29] We can explain the automatic and second-nature character of this performance-conditional equivalence by appeal to a folk theory that is well-entrenched in our ways of thinking about thought and speech. On that folk theory, thoughts are the kind of state that can be expressed in speech; so anybody who says that it's raining will have the thought that it's raining. Our operating on that tacit assumption leads to our regarding as indifferent whether one says that it's raining or that one thinks it's raining.

[30] There is an exception. Saying that I think it's raining verbally expresses a HOT that I think it's raining but, as noted earlier, that HOT is seldom conscious. But the HOT model also explains this exception, since the higher-order case differs in a crucial way from the first-order case. Though the statement that I think it's raining is, again, performance conditionally equivalent to the statement that I think that I think that it's raining, in this case the performance-conditional equivalence is hardly second nature. So we needn't suppose that any third-order thought occurs.

[31] For more on this, see Rosenthal (Why Are Verbally Expressed Thoughts Conscious?, forthcoming).

[32] Freud's well-known example in *The Psychopathology of Everyday Life*

as expressing the person's unconscious intention to close the meeting. And perhaps, then, that the speech act expresses a thought that isn't, after all, conscious.

But that construal doesn't do the situation full justice. The person also has some conscious intention to open the meeting; otherwise why was any remark made at all? And, if we see the remark as expressing the unconscious intention not to open the meeting, the connection between the conscious intention and the making of the remark remains obscure. We would do better to see the remark as verbally expressing the conscious intention, and regard the unconscious intention not as expressed by the remark but as interfering with its production, by causing the injection of a stray word. Parapraxis is an interaction effect in which an unconscious state affects the expressing of a conscious state; it is not a counterexample to the regularity that verbally expressed cognitive states are invariably conscious.[33]

Things are strikingly different with affective states, such as joy, anger, and sadness. One might verbally express anger, for example, by asserting moral disapprobation or some other negative evaluation. And in that case one may well also adopt a kind of emotional distance from the situation, even to the point of remaining unaware of one's anger. This can happen because, unlike the case of cognitive intentional states, there is no standard way of verbally expressing one's affective states. So there is no well-entrenched equivalence in conditions of use between one's verbally expressing an affective state and one's describing oneself as being in that state, and hence no reason to posit a HOT whenever one verbally expresses an affective state. An affective state's being verbally expressed is not by itself sufficient for that state to be conscious.[34]

As with all other mental states, however, being able to report an affective state noninferentially is enough for that state to be conscious, since that ability ensures the occurrence of the HOT that an actual report would express. And a noninferential report expresses one's consciousness of oneself as being in the state in question.

(Freud 1901b, p. 59-60), in which the President of the Lower House of the Austrian Parliament's said "I . . . declare the sitting *closed*," rather than open. See also Freud (1940b) and elsewhere.

[33] It might also be that both intentions occur consciously, though for social reasons the speaker would resist acknowledging the intention not to open the meeting. In such a case, even if the remark expressed the intention not to open the meeting, it would still express a conscious intention.

[34] On the special case of affective states, see Rosenthal (1998).

A mental state's being conscious consists in one's being conscious of that state in some suitable way. Inner-sense theorists typically construe our being thus conscious of our mental states as the result of some mechanism by which the mind monitors itself (see, e.g., Armstrong 1980; Lycan 1996). Mechanisms can malfunction, and a monitoring mechanism could deliver mistaken results; so the existence of some such mechanism would square with the observation that consciousness is neither infallible nor exhaustive. But consciousness doesn't simply go wrong; it sometimes actually confabulates. And that suggests that self-interpretation plays some role in whatever process leads to our being noninferentially conscious of our mental states. This, once again, points to HOTs rather than inner sense.

Even independent of confabulation, it is inviting to see noninferential reports that one is in some mental state as selfinterpretations; one interprets oneself as being in the mental state in question. Others sometimes offer interpretations of what mental state one is in; but, as noted earlier, accepting another's interpretation does not by itself lead to the state's being conscious. Accepting another's interpretation of one's mental state is coming to think that one is in that state. But that thought will remain inferential as long as it rests on one's accepting the other person's interpretation.

But if one comes to identify with that other person's interpretation sufficiently closely to make that interpretation one's own, the resulting self-interpretation may well become noninferential. One will then be conscious of oneself as being in the state, but it will no longer seem to one as though that awareness of the state relies on any conscious inference or other mediating process. One will seem subjectively to be directly aware of the state, and so that state will have come to be conscious. As mentioned earlier, one need not also cease being aware of one's having previously relied on the other person's interpretation; it's enough that one now takes oneself to be in the relevant state independently of any such inference.

What happens here resembles the process by which all verbally expressed thoughts come to be conscious. In both cases, the performance of a speech act leads automatically and without conscious processing to one's having a HOT that one is in a particular mental state, and so to one's being conscious of oneself as being in that state. The ascription by the other person results in one's having a HOT because one identifies with that person and, hence, with the ascription. One's verbally expressing a thought leads to one's having a HOT about it because of the well-entrenched equivalence between saying something and saying that one thinks that thing. And we can see that automatic equivalence, in turn, as itself a matter of identifying with one's own speech act.[35]

Having noted that we often infer what mental states others are in, Freud writes that "[p]sycho-analysis demands nothing more than that we should apply this process of inference to ourselves also" (Freud 1915e, p. 169). But, if that process of inference about ourselves is to lead to the inferred mental states' being conscious, it must come to seem as though that inference no longer figures in our being conscious of ourselves as being in those states.[36]

REFERENCES

Aristotle De Anima. In Aristotle´s De Anima. Books II and III. Oxford: Clarendon Press 1993.

Armstrong, D.M. (1980). What is consciousness. In The Nature of Mind. St. Lucia, Queensland: University of Queensland Press, pp. 55–67.

Baars, B.J. (1988). A Cognitive Theory of Consciousness. Cambridge: Cambridge University Press.

Baars, B.J. (1997). In the Theater of Consciousness: The Workspace of the Mind. New York: Oxford University Press.

Block, N. (1986). Advertisement for a semantics for psychology. Midwest Studies in Philosophy X: 615–678.

Block, N. (1995). On a confusion about a function of consciousness. The Behavioral and Brain Sciences 18(2): 227–247.

Brentano, F. (1884). Psychology from an Empirical Standpoint. London: Routledge & Kegan Paul 1973.

Broad, C.D. (1925). The Mind and its Place in Nature. London: Routledge and Kegan Paul.

Carruthers, P. (2000). Phenomenal Consciousness: A Naturalistic Theory. Cambridge: Cambridge University Press.

Dennett, D. (1991). Consciousness Explained. Boston: Little, Brown and Company.

---

[35] One identifies with one's own speech act by way of the tacit, well-entrenched folk theory on which speech acts all express intentional states. On that folk theory, see n. 29, above. One's nonconscious thought that one thinks the thing one is saying constitutes, in effect, an identification of oneself with the thought one takes oneself to have.

[36] A first version of this paper was presented at the 3rd International Symposium of the Austrian Academy of Sciences "Psychoanalysis as an Empirical, Interdisciplinary Science. On Continental & Anglo-American Research", Vienna, November 22, 2002. I am very grateful to that audience for useful reactions.

Dennett, D. (1993). The message is: There is no medium. Philosophy and Phenomenological Research 53(4): 919–931.

Descartes, R. (1641) Geometrical Exposition of the Second Replies. Oeuvres de Descartes 7, ed. C. Adam & P. Tannery. Paris: J. Vrin 1964.

Descartes, R. (1645) Letter to Newcastle. Oeuvres de Descartes 4, ed. C. Adam & P. Tannery. Paris: J. Vrin 1964, pp. 573–576.

Descartes, R. (1649) Letter to More. Oeuvres de Descartes 5, ed. C. Adam & P. Tannery. Paris: J. Vrin 1964, pp. 275–279.

Fodor, J.A. (1992). A theory of the child's theory of mind. Cognition 44(3): 283–296.

Freud, S. (1900a). The Interpretation of Dreams. S.E. 4–5.

Freud, S. (1901b). The Psychopathology of Everyday Life. S. E. 6: 1–289.

Freud, S. (1915e). The Unconscious. S. E. 14: 166–215.

Freud, S. (1923b). The Ego and the Id. S. E. 19: 3–68.

Freud, S. (1940a). An Outline of Psychoanalysis. S. E. 23: 141–207.

Freud, S. (1940b). Some elementary lessons in psycho-analysis. S. E. 23: 279–86.

Gennaro, R.J. (1995). Consciousness and Self-Consciousness. Amsterdam and Philadelphia: John Benjamins Publishing Company.

Greenwald, A.G. & Banaji, M.R. (1995). Implicit social cognition: Attitudes, self-esteem, and stereotypes. Psychological Review 102(1): 4–27.

Harman, G. (1973). Thought. Princeton: Princeton University Press.

Harman, G. (1999). Reasoning, Meaning and Mind. Oxford: Clarendon Press.

Holmes, D.S. & Frost, R.O. (1976). Effect of false autonomic feedback on self-reported anxiety, pain perception, and pulse rate. Behavior Therapy 7(3): 330–334.

Kant, I. (1787). Critique of Pure Reason. Cambridge: Cambridge University Press 1998.

Kriegel, U. (2002). Consciousness, permanent self-awareness, and higher-order monitoring. Dialogue 41(3): 517–540.

Kriegel, U. (2003). Consciousness, higher-order content, and the individuation of vehicles. Synthèse 134: 477–504.

Locke, J. (1700). An Essay Concerning Human Understanding. Oxford: Oxford University Press 1975.

Lycan, W.G. (1996). Consciousness and Experience. Cambridge, Mass.: MIT Press/Bradford Books, pp. 13–43.

Marcel, A.J. (1983a). Conscious and unconscious perception: Experiments on visual masking and word recognition. Cognitive Psychology 15: 197–237.

Marcel, A.J. (1983b). Conscious and unconscious perception: An approach to the relations between phenomenal experience and perceptual processes. Cognitive Psychology 15: 238–300.

Merikle, P.M., Smilek, D. & Eastwood, J.D. (2001). Perception without awareness: Perspectives from cognitive psychology. Cognition 79(1–2): 115–134.

Moore, G.E. (1942). A reply to my critics. In The Philosophy of G. E. Moore, ed. P.A. Schilpp. New York: Tudor Publishing Company, pp. 533–677.

Moore, G.E. (1944). Russell's 'theory of descriptions'. In The Philosophy of Bertrand Russell, ed. P.A. Schilpp. New York: Tudor Publishing Company, pp. 175–226.

Nisbett, R.E. & DeCamp Wilson, T. (1977). Telling more than we can know: Verbal reports on mental processes. Psychological Review 84(3): 231–259.

Perner, J., Leekam, S.R. & Wimmer, H. (1987). Three-year olds' difficulty with false belief: The case for a conceptual deficit. British Journal of Developmental Psychology 5(2): 125–137.

Perner, J. (1991). Understanding the Representational Mind. Cambridge, Mass.: MIT Press/Bradford Books.

Perner, J. (1995). The many faces of belief: Reflections on Fodor's and the child's theory of mind. Cognition 57(3): 241–269.

Rosenthal, D.M. (1986). Two concepts of consciousness. Philosophical Studies 49(3): 329–359.

Rosenthal, D.M. (1993). Thinking that one thinks. In Consciousness: Psychological and Philosophical Essays, ed. M. Davies and G.W. Humphreys. Oxford: Basil Blackwell, pp. 197–223.

Rosenthal, D.M. (1995). Moore's paradox and consciousness. Philosophical Perspectives 9: 313–333.

Rosenthal, D.M. (1997a). Perceptual and cognitive models of consciousness. Journal of the American Psychoanalytic Association 45(3): 740–746.

Rosenthal, D.M. (1997b). A theory of consciousness. In The Nature of Consciousness: Philosophical Debates, ed. N. Block, O. Flanagan & G. Güzeldere. Cambridge, Mass.: MIT Press, pp. 729–753.

Rosenthal, D.M. (1998). Consciousness and its expression. Midwest Studies in Philosophy 22: 294–309.

Rosenthal, D.M. (1999a). The colors and shapes of visual experiences. In Consciousness and Intentionality: Models and Modalities of Attribution, ed. D. Fisette. Dordrecht: Kluwer Academic Publishers, pp. 95–118.

Rosenthal, D.M. (1999b). Sensory quality and the relocation story. Philosophical Topics 26(1–2): 321–350.

Rosenthal, D.M. (2002). Explaining consciousness. In Philosophy of Mind: Classical and Contemporary Readings, ed. David J. Chalmers. New York: Oxford University Press, pp. 406–421.

Rosenthal, D.M. (2003). Unity of consciousness and the self. Proceedings of the Aristotelian Society 103(3): 325–352.

Rosenthal, D.M. (2005). Why are verbally expressed thoughts conscious? In Consciousness and Mind, ed. D.M. Rosenthal. Oxford: Clarendon Press, in press.

Ryle, G. (1949). The Concept of Mind. London: Hutchinson and Company.

Solms, M. (1997a). What is consciousness? Journal of the American Psychoanalytic Association 45(3): 681–703.

Solms, M. (1997b). Response. Journal of the American Psychoanalytic Association 45(3): 765–778.

Staats, P.S., Hekmat, H. & Staats, A.W. (1998). Suggestion/placebo effects on pain: Negative as well as positive. Journal of Pain and Symptom Management 15(4): 235–243.

Weiskrantz, L. (1986). Blindsight: A Case Study and Implications. Oxford: Clarendon Press.

Weiskrantz, L. (1997). Consciousness Lost and Found: A Neuropsychological Exploration. Oxford: Clarendon Press.

Wellman, H.M. (1990). A Child's Theory of Mind. Cambridge, Mass.: MIT Press/Bradford Books.

White, P.A. (1988). Knowing more than we can tell: 'Introspective access' and causal report accuracy 10 years later. British Journal of Psychology 79(1): 13–45.

Wilson, T.D., Hodges, S.D. & LaFleur, S.J. (1995). Effects of introspecting about reasons: Inferring attitudes from accessible thoughts. Journal of Personality and Social Psychology 69(1): 16–28.

Wimmer, H. & Perner, J. (1983). Beliefs about beliefs: Representation and constraining function of wrong beliefs in young children's understanding of deception. Cognition 13(1): 103–128.

Wittgenstein, L. (1953). Philosophical Investigations, ed. G. E. M. Anscombe. New York: Macmillan.

VI. FIRST-PERSON AND THIRD-PERSON MINDREADING

PIERRE JACOB

1. INTRODUCTION

My topic is what the developmental cognitive psychologist Simon Baron-Cohen (1995) aptly calls "mindreading". As the label suggests, mindreading is the human ability to recognize the presence of minds. Full-blown human mindreading culminates in the practice of representing and ascribing states of mind – states of minds of others and states of one's own mind. Typically, human beings ascribe emotions and so-called propositional attitudes (such as intentions, beliefs and desires) to other agents for the purpose of explaining and predicting their intentional actions. A person represents her own beliefs, intentions and desires when e.g., she is in the process of revising her goals and/or her beliefs and is checking the consistency of her own mind. Alteration of mindreading can result in "mindblindness". As Baron-Cohen (1995) has put it, "imagine what your world would be like if you were aware of physical things but were (mind-) blind to the existence of mental things [...] blind to things like thoughts, beliefs, knowledge, desires, and intentions, which for most of us self-evidently underlie behavior."

If and when it is reliable, first-person mindreading – traditionally called "introspection" – gives rise to self-knowledge or true beliefs about the contents of one's own mind. Third-person mindreading gives rise to knowledge of, or true beliefs about, the contents of other minds. From an evolutionary standpoint, the adaptive benefits of third-person mindreading among humans are quite obvious. If one individual is contemplating the choice between cooperation and competition with another agent, then it is useful for the former to be able to detect accurately the latter's goals and intentions and vice-versa. Arguably, if one did not know the content of one's *own* mind, then one could not even contemplate the choice between cooperation and competition. If so, then the adaptive advantages of both third-person and first-person mindreading stand or fall together. Plausibly then, they must have co-evolved in tandem.

The goal of my paper is to assess a recent proposal in the cognitive sciences and in the philosophy of mind according to which all aspects of mindreading can be accounted for by a simple cognitive heuristic called "mental simulation". I will argue that some fundamental aspects of full-blown third-person mindreading lie outside the scope of mental simulation and that mental simulation fails to account for first-person mindreading. Before I proceed, however, I will first make a couple of assumptions explicit and then I will make three disclaimers.

2. A PAIR OF ASSUMPTIONS

Human beings are self-conscious and they are conscious of others. I assume that both one's awareness of others and one's self-awareness can be divided into awareness of two kinds of properties: awareness of non-psychological or bodily properties and awareness of psychological properties. There are two broad ways in which one can become aware of both the bodily properties of others and one's own bodily properties.

On the one hand, one can use one's sense organs to perceive the non-psychological properties of others. One can also see, hear and smell some – not all – of one's bodily parts. One can hear one's own voice. One can have proprioceptive information about the relative positions of one's limbs and about one's posture. Indeed, as Gibson (1979) insisted and as contemporary cognitive neuroscience confirms, the visual system could not represent some of the properties of distal stimuli unless it took into account information about some of the observer's movements. For example, the visual computation of the position of an object relative to an agent is based on information about the agent's eye- and head-movements. Thus, visual percepts of non-mental objects may depend upon the processing of non-perceptual information about the bodily self or what Bermudez (1998) calls "the embodied self".

On the other hand, one can also come to learn facts about human anatomy and human physiology, and thus about one's bodily properties, not by perceiving bodily properties – neither the bodily properties of others, nor one's own bodily properties –, but by reading books about them. By reading books, one can come to know anatomical and physiological facts about e.g., the human skeleton, the human digestive system, the human cardiovascular system, the human immune system or the human nervous system. In so doing, one becomes *conceptually* aware of the bodily (or non-psychological) properties of both oneself and others.

Psychological self-knowledge (or first-person mindreading) raises two questions. First, I can be aware both of my non-psychological properties and of my psychological properties. Given that I can be aware of my non-psychological properties in two ways, the first question is whether my awareness of my own psychological properties comes in two varieties too. Can one be both perceptually and conceptually aware of one's own psychological properties? Secondly, I can be aware of both your psychological properties and my psychological properties. I can do third-person mindreading and first-person mindreading. It is obvious that I must rely on the observation of your behavior in order to know your mind. What about first-person mindreading? Do I also need to observe my own behavior to do first-person mindreading?

The word "introspection" refers to the process whereby one person is aware of her own psychological properties. Many philosophers – especially in the classical empiricist tradition – have assumed that one can be perceptually aware of one's own psychological properties. I think that this is a deep mistake. Observation of the behavior of others is required to know their psychological properties. One can be perceptually aware of some of one's non-psychological properties. But awareness of one's own psychological properties can only be conceptual. Introspection is a conceptual, not a perceptual, mechanism (as I will argue). On my view, Shoemaker's (1968) well-known claim that one's knowledge of oneself is immune to errors of misidentification is restricted to introspective knowledge of one's own psychological properties.[1]

3. THREE DISCLAIMERS

My first disclaimer has to do with the scope of the concept of *knowledge* as I understand it here for the purpose of analyzing how humans come to know the contents of either their own minds or the minds of others. Philosophers – and especially contemporary analytic epistemologists – are concerned with a restricted sense of the English verb 'to know'. In this restricted sense, a person can be said to know proposition *p* only if she believes that proposition *p* is true, proposition *p is* indeed true and some further condition is satisfied so that the person's belief has not been formed out of shere luck or mere fantasy. According to externalist epistemologists, the belief must have arisen by virtue of some reliable mechanism. According to internalist epistemologists, a person will fail to know that *p* unless she has a true justified higher-order belief

[1] A claim anticipated by Wittgenstein and explored in his own way by Evans (1982).

that her true first-order belief is itself justified. In this paper, I do not use the verb "to know" in this restricted epistemological sense. Like psychologists, if and when I say that person *X* knows either that she believes so and so or that someone else desires such and such, I do not mean that *X*'s knowledge qualifies as "knowledge" in the epistemologist's favored sense.

Secondly, one should not expect from the cognitive scientific study of third-person mindreading a response to the challenge of philosophical *skepticism* about knowledge of other minds. Philosophical skeptics worry about whether knowledge e.g., of the external world is possible *at all*.[2] Suppose I claim that I know that my neighbor Ann is at home because I have seen her car parked in front of her house. As a matter of fact, Ann's car would not be parked in front of her house unless she was at home and I know it. Now, seeing Ann's car parked in front of her house may well be a reliable method for knowing whether she is at home. But the philosophical skeptic will not be satisfied, because the explanation of how I know something about Ann (namely that she is at home) takes for granted that I know something else about her car (namely that it is parked in front of her house). And the skeptic refuses to take any knowledge for granted.

Similarly, in the case of knowledge of other minds. Suppose I claim to know what Ann thinks because either I asked her or I asked her husband. Asking her or asking her husband may be a reliable method for knowing what Ann thinks on a particular topic. However, if this is the way I come to know what Ann thinks, then I take for granted either that she knows what she thinks or that her husband does. Therefore, I fail to satisfy the skeptic's requirement for a perfectly general account of knowledge of what someone thinks – i.e., an account that does not presuppose any other knowledge. So my second disclaimer is that we should not expect the cognitive scientific study of third-person mindreading to meet the challenge of skepticism about knowledge of other minds. Nor should we expect it to offer a perfectly general account of knowledge of other minds. Rather, we should expect it to tell us something about the actual mechanisms underlying our knowledge of other minds.

Thirdly, the idea that *introspective* beliefs about facts involving one's own psychological properties (such as believing so-and-so or experiencing such-and-such) enjoy a unique *epistemic authority* or privi-

---

[2] See Stroud (1989).

lege has played a fundamental though different role in traditional rationalist epistemology and in traditional empiricist epistemology. In rationalist epistemology, the primary target of introspective self-knowledge are one's own thoughts. In empiricist epistemology, the primary target of introspective self-knowledge are one's own sense-data or perceptual experiences. I do not think that the traditional epistemological privileges ascribed to introspective self-knowledge are justified. I do not take for granted either that self-knowledge is a priori knowledge or that it is infallible.

On the one hand, in rationalist epistemology, psychological self-knowledge was taken to be one of two paradigms of a priori human knowledge, the other paradigm being knowledge of mathematical truths. Rationalist epistemology has three ingredients. First, it is of the essence of the Cartesian mind that it entertains or forms thoughts (as opposed to having e.g., perceptual experiences). Secondly, thoughts have concepts or ideas as constituents. Thirdly, the mind is transparent to itself: one cannot have a thought of which one is not aware. In rationalist epistemology, what secures both the priority and infallibility of one's introspective awareness of one's own psychological properties is that all the psychological properties a mind can exemplify are properties of thoughts or judgments, not experiences. Given that the mind is transparent to itself or that thoughts are reflexive in the sense (of the transparency assumption) that one cannot entertain a thought (or make a judgment) unless one is aware that one is, it follows that introspective knowledge of one's own mind is a priori and, according to Descartes, infallible.

In traditional empiricist epistemology, on the other hand, the most primitive and elementary constituents of minds are perceptual experiences or sense-data, not concepts. Concepts (of either psychological or non-psychological properties) are logical constructions out of sense-data. According to much traditional empiricist epistemology from Locke to Russell, knowledge of the external world – knowledge of mind-independent facts – is twice dependent on psychological self-knowledge. First of all, knowledge of mind-independent facts depends on the epistemologically antecedent knowledge of mental or psychological facts about oneself (such as that one is having a particular perceptual experience or sense-datum). Secondly, one's knowledge of mental or psychological facts about oneself derives in turn from one's direct quasi-perceptual acquaintance with some mental entity present to or in one's mind, i.e., the sense-datum or perceptual experience. If one's awareness of one's sense-data consists in being acquainted with them, then one is made aware of one's sense-data by some kind of quasi-perceptual process or peering inside at one's own perceptual experiences.

Both the rationalist and the empiricist pictures of introspective self-knowledge have come under serious criticism in contemporary philosophy and psychology. On the one hand, the empiricist model of a quasi-perceptual process whereby one becomes self-aware of one's own perceptual experiences has been systematically dismantled. For detailed criticism of the quasi-perceptual model of psychological self-knowledge, see the last section of this paper.

On the other hand, the Cartesian picture of introspection raises at least two issues. First, the asymmetry between first-person and third-person mindreading that results from a Cartesian picture of introspective self-knowledge raises a genuine puzzle. As Davidson (1984, 1987) recognizes, the asymmetry between first-person and third-person mindreading takes it for granted both that claims to know one's own mind are made independently of any empirical evidence and that they enjoy an epistemic authority of which third-person claims to know the minds of others are deprived. The puzzle is: why should claims *without* evidential support have more epistemic authority than claims based on evidence? The second question is: given that the Cartesian assumption that the mind is transparent to itself has come under heavy attack, what is left of the Cartesian picture of the special epistemic authority of introspective beliefs? Since Freud, it is commonly accepted that one may be blind to some of one's own beliefs and desires. You may be better positioned than I am to get insight into some of my beliefs and desires. Furthermore, a human mind does not merely entertain thoughts and propositional attitudes; it also has perceptual experiences. On the Cartesian picture, the immunity to error of the mind's introspective beliefs about itself was secured by the joint assumptions that it is of the essence of the mind to entertain thoughts and that one cannot entertain a thought unless one knows that one is doing so. But how could such assumptions entail that one's introspective beliefs are exhaustive and exhaustively true a priori? How could they secure a priori true introspective beliefs about the content and attitude of each of one's own propositional attitudes – both one's occurrent propositional attitudes and one's dispositional propositional attitudes? Furthermore, how could such assumptions secure introspective a priori knowledge about one's perceptual experiences at all? At the minimum, the Cartesian picture of self-knowledge would need to assume that perceptual experiences are judgments (or that the phenomenal character of a perceptual experience is derivative from a judgment)[3].

4. WHAT IS MENTAL SIMULATION?

The idea of mental simulation involves three complementary ingredients. First, the idea of mental simulation as a basis for mindreading derives from the rejection of the picture of mindreading as a capacity for *detached* theorizing. Secondly, much of the appeal of mental simulation derives from the fact that it is at once a simple and a general-purpose, not a domain-specific, mental heuristic. As such, it holds the promise of unifying various aspects of human cognition. The idea is that any human cognitive mechanism can be used "off-line" in a "pretend" mode. Mental simulation *is* this pretend mode. Finally, proponents of the application of the mental simulation strategy to mindreading assume that the fundamental goal of mindreading is mental mimicry or the duplication of someone's mental life. The idea that mental simulation has the potential to account for the human mindreading capacity has arisen in two scientific contexts: in the context of the investigation by philosophers and developmental cognitive psychologists of the human capacity to do mindreading and in the context of the cognitive neuroscientific study of motor cognition.

The simulation approach to mindreading derives some of its appeal from some of the shortcomings of some of the rival approaches to mindreading which have emerged in both philosophy and cognitive science in the last twenty years or so. One such influential rival approach is the so-called "theory-theory" (or TT) approach to mindreading (associated with the work of Gopnik, Meltzoff and Wellman).[4] This approach has arisen in response to two fundamental questions: 1. What is the fundamental structure of the human adult ability to mindread? 2. How is it acquired both from an ontogenetic and a phylogenetic point of view? According to TT approaches, the response to the first question is that human adults tacitly know a naive psychological theory comprised of naive psychological laws ranging over abstract unobservable mental states such as intentions, beliefs and desires. The response to the ontogenetic part of the second question is that human infants are "little sci-

---

[3] A view that is highly questionable.

[4] See Gopnik & Wellman (1992, 1994) and Gopnik & Meltzoff (1997).

entists": they use their science-forming capacity to discover the laws of naive psychology. In this respect, normal ontogenetic development undergoes a series of theory-changes pretty much like those exhibited in the history of scientific theorizing.

Some of the implausible features of TT views have prompted the emergence of an alternative nativist modular domain-specific approach. On this alternative modular view (associated with the work of Baron-Cohen and Leslie),[5] mindreading results from a set of distinct special purpose modules.[6] Each such module has a dedicated task: ID (intentionality detector), for example, takes patterns of motion of an object and ascribes goal or agency to the moving object. EDD (eye-direction detector) takes movements of the eyes in their orbits as input and finds some object as the target of the agent's gaze-direction. SAM, which is a shared or joint attention mechanism, takes as input the output of EDD and represents some object as the target of the joint attention of at least two agents (one of which can be the self). ToMM is the theory of mind mechanism proper, whose job is to deliver ascriptions of beliefs and desires to an agent.[7] With an eye on the phylogenetic part of the second question, advocates of the nativist modular view entertain the hypothesis that several such modules may be present in the brain of non-human primates. In the brain of human infants, the outputs of ID and EDD serve as inputs to SAM whose output serves as input to ToMM. In the course of human development, this chain of modules unfolds according to a strict ontogenetic schedule.

Both theory-theorists and nativist modular theorists derive inspiration from the Chomskyan study of the language faculty where the ability of a speaker of a natural language to produce and understand an infinite set of grammatical strings of words of her language is explained in terms of her detached tacit knowledge of the grammar of the language. This is what simulation theorists reject and this is the first component of the mental simulation approach to mindreading. According to simulation theorists, mindreading reflects no so much a capacity for detached theorizing about mental states as a practical ability to *do* something, i.e., to

[5] See Baron-Cohen (1995), Leslie (1994) and Leslie and Thaiss (1992).

[6] I take Carey & Spelke's (1994) and Spelke's (2000) favored notion of "core knowledge" to be more in line with what I call the "nativist modular domain-specific" approach to human cognitive development than with the "theory-theory" approach.

[7] See in particular Baron-Cohen (1995).

engage into a certain kind of *action*, namely the impersonation or the replication of the mental lives of one's conspecifics.[8] This is, I take it, what Gordon (1996, p. 11), a simulation theorist, has in mind, when he contrasts what he calls a "hot methodology" (one "which exploits one's own motivational and emotional resources") and a "cold methodology" (i.e., a "methodology that chiefly engages our intellectual processes").

The second and third features of the mental simulation approach to mindreading derive from the neuroscientific investigation of the motor cortex involved in the planning, the execution and the perception of human actions. The neuroscientific investigation of motor cognition in human and non-human primates has revealed that many of the brain mechanisms involved in performing particular actions are also involved in the mental preparation of (or in imagining) such actions and in the perception of such actions performed by others. On the one hand, most of the neural events whose unfolding gives rise to an *overt* (executed) action are involved either in its *covert* stages or in tasks of *motor imagery* in which the subject is required to imagine an action without executing it. On the other hand, most of the neural events involved in the covert stage of an action or in tasks of motor imagery are involved in the perception of a different token of the same type of action performed by another.[9]

One of the most startling findings of the cognitive neuroscience of motor cognition has been the discovery by Di Pellegrino et al. (1992), Rizzolatti et al. (1995), in area F5 of the pre-motor cortex of macaque monkeys, of a population of neurons – so-called 'mirror neurons' – that discharge selectively both when the animal performs some very specific object-directed hand movements and when the animal visually perceives the same kinds of object-directed movements performed by a conspecific. For example, mirror neurons have been found to discharge selectively for the execution and the perception of hand movements involved in holding, grasping, manipulating or tearing raisins or peanuts.

On the basis of such remarkable findings, simulation theorists have offered the suggestion that something like the mirror system in humans might be what Gallese (2003) calls a "shared manifold of interpersonal relations" (or social cognition). The idea is that mirror neurons offer a

[8] I borrow the contrast between engaged (or immersed) cognitive activity and detached theorizing to Campbell (1994), who applied it to naive thinking about space and time.

[9] Cf. Jeannerod (1997, 2001, 2002).

common neural basis not only for the execution and the preparation (or the imagining) of one's own actions, but also for the understanding of the actions of others. The idea that mental simulation is a cognitive heuristic underlying all forms of mindreading can be seen as a generalization of the discovery that in motor cognition, one and the same neural structure underlies the control and the imagination of one's own actions as well as the perception of the actions of others.

Any cognitive mechanism that takes standard inputs and produces standard outputs in response can be used *off-line*, i.e., it can be given "pretend" inputs and produce "pretend" output in response. For example, Currie (1995) argues that visual mental imagery is the off-line use of the visual system whereby it processes inputs from memory rather than retinal images. In imagining an action or in perceiving an action performed by another, one's motor motor system is used *off-line.* In the same vein, Stich & Nichols (1992, 1995) have hypothesized that one's practical reasoning (or decision) system can be used "off-line" in order to predict someone else's decisions. When used on-line, one's decision system takes one's beliefs and desires as inputs and produces decisions on the basis of which one acts. When used off-line, the same mechanism takes "pretend" beliefs and desires as input and produces a "pretend" decision, which is then used as a prediction of the other's decision. The hypothesis that any cognitive mechanism can be used off-line is the second ingredient of mental simulation.

The third ingredient of the idea of mental simulation is mental mimicry. As Gallese & Goldman (1998, p. 495) have emphasized, the motor system seems to "form a cortical system that matches observation and execution of motor actions." It is tempting – and indeed it has tempted many cognitive scientists and some philosophers as well – to suppose that the main function of the mirror system in both humans and nonhuman primates is to enable imitation. If, as most simulation theorists assume, the major function of the human mindreading system in turn consists in mental imitation (or in mimicking the mental life of a conspecific), then it will seem natural to assume that the mirror system underlies the human mindreading system. Both the view that the mirror system is a sufficient condition of human imitation and the view that the main goal of the human mindreading system is to reproduce the mental life of one's conspecific seem to me deeply mistaken. I shall start with the first mistake.[10]

[10] Note that the idea that the function of the mindreading system is to imitate

The view that the mirror system is a sufficient basis for human imitation embodies, I think, two mistakes. The first mistake is the conceptual confusion between *imitation* and *mimicry*. Imitation is very widespread in humans, but as Tomasello (2000) and Hauser et al. (2002) argue, monkeys and apes do *not* imitate very much, if at all. Spontaneous cross-modal imitation seems sharply limited to humans. There are mirror neurons in the premotor cortex of macaque monkeys, but macaque monkeys do not engage in imitation. By contrast, as observed by Meltzoff and Moore (1977, 1997), human infants only a few hours old are able to reproduce facial gestures that they have seen. Human infants are not bound to the mere reproduction (or copy) of bodily movements. Very early on, they can retrieve the intention underlying an action and perform a successful imitation of a failed exemplar of an action (Meltzoff, 1995). Human adults can imitate pretty much anything: human imitation ranges from non-intentional noises or movements to the intentional behavior of conspecifics. When imitating the intentional behavior of a conspecific, humans do not merely copy bodily gestures. They retrieve the intention underlying the imitated behavior.

Among members of non-human species including non-human primates, one finds ubiquitous instances of *mimicry*, not imitation. Whereas imitation is, mimicry is *not*, intentional behavior. Unlike human imitation, which may require the retrieveal of the intention underlying the observed piece of behavior, mimicry is a process whereby A's perception of some property exemplified in the environment automatically causes A to exemplify the property in question. For example, the monarchs are butterflies that are preys of birds (the bluejays). The former store and secrete a substance toxic to their predators. Other butterflies – the vice-roys – do not secrete the toxic substance, but instead they mimick the colors of the monarchs, thereby misleading the bluejays into 'believing' that they too are toxic. Mimicry is ubiquitous in non-human animals and in human behavior as well. Thus, emotional contagion, laughter and yawning are cases of mimicry whereby A's perception of B's behavior automatically (or non-intentionally) prompts A to reproduce B's behavior. But they are not instances of imitation.[11]

The second mistake is the assumption that the mirror system fires in response to the perception of social actions. Human and non-human

---

makes sense when applied to third-person mindreading, but not to first-person mindreading.

[11] Cf. Jacob & Jeannerod (2003, ch. 7).

primates can perceive the actions of their conspecifics. Some of the perceived actions are themselves directed towards conspecifics. Others are directed towards other things. Only actions directed towards conspecifics (e.g., communication) are distinctly social actions and only social actions are guided by social intentions, i.e., intentions to modify a conspecific's behavior.

When mirror neurons fire in the brain of monkey A observing the performance of an action executed by monkey B, B presumably intends to act on an object, but B does *not* intend ipso facto to modify A's behavior. B's intention is a *motor*, *not* a *social*, intention. The role of the mirror system in primates seems interestingly limited to matching execution and observation of *object*-oriented actions. Thus, mirror neurons are involved in the process of retrieving from perceptual cues motor intentions involved in an agent's object-directed actions. By contrast, neurons have been registered by Perrett et al. (1982, 1990) in a brain area called the superior temporal sulcus (STS), which is near but distinct from area F5 where the activity of mirror neurons has been recorded. Neurons in STS fire in response to the perception of movements of bodily parts that are not restricted to object-oriented actions and that are likely to convey emotional and social information about the agent. For example, they discharge upon seeing eye-gaze direction, movements of the lips, upwards and downwards head-movements, movements of the hand. But, as Rizzolatti et al. (2000, p. 548) emphasize, unlike mirror neurons, neurons in STS are purely perceptual, not sensori-motor, neurons. Thus, the most plausible interpretation of the mirror system in monkeys is that it constitutes an engaged proto-conceptual system designed for the recognition of object-oriented actions. It is engaged in the sense that it will not recognize an action unless the neural resources required for an animal to prepare the execution of that very action are active. Its recognitional powers are limited to object-oriented actions and motor intentions.

5. MENTAL SIMULATION AND THIRD-PERSON MINDREADING

There are two versions of the theory of mental simulation as it applies to third-person mindreading: Goldman's (1989, 1993, 2000) introspectionist version and Gordon's (1986, 1995, 1996) anti-introspectionist version. According to the former, knowledge of other minds derives from knowledge of one's own mind. In other words, I use knowledge of my own mind as a model from which I inferentially construct knowledge of other minds. The introspectionist model of third-person mindreading is thus confronted to the preliminary question: how do I derive knowledge of my own mind? I will examine this question in the next and final section. I presently turn to the latter anti-introspectionist version of the simulation theory, according to which third-person mindreading consists in *pretence*. Both versions of the simulation theory, however, agree that the fundamental goal of third-person mindreading is to duplicate, reproduce or mimick the mental life of another (the third feature of the idea of mental simulation). Before criticizing the anti-introspectionist approach to third-person mindreading, let me say what is wrong with the third feature of the idea of mental simulation as applied to third-person mindreading.

The goal of mindreading is to form an accurate picture of what our conspecifics think and have in mind. As theory-theorists have, I think, rightly emphasized, what other people *do* might be observable. At least, their bodily movements are. But what they have in mind – their intentions, beliefs, desires, feelings and emotions – is not directly observable in their actions or behavior. Whether or not the mental simulation (pretence) heuristic is a fundamental part of our method for accomplishing third-person mindreading, our success in the task does presuppose a certain amount of psychological similarity between members of the human species. As simulation theorists have emphasized, existing mechanisms of mimicry may serve a useful purpose: they may enhance the probability of success in tasks of understanding the minds of one's conspecifics. For reasons that will emerge soon, neither mimircy nor even imitation, however, can be the process of third-person mindreading.

Why cannot mindreading be mental imitation, let alone mental mimicry? When I try to understand one of your mental states, e.g., one of your beliefs, I try to capture two features of your mental state: its *content* and the *attitude* under which you entertain the content. The only way I can capture the attitude is by applying a mental concept (i.e., the concept of *belief*). The only way I can capture the content is by exploiting a different representation whose content will be suitably related to the content of your belief. Neither the attitude nor the content of a mental state can be copied.

Genes can be copied because, unless they mutate, they replicate. Utterances of sentences of natural and other languages can be copied or reproduced in virtue of their phonological non-semantic properties. Arguably, direct quotation and photocopying are methods for reproducing utterances. But when it comes to an appreciation of the content or semantic properties of utterances, reproduction or copying is not an option anymore. Only *interpretation* is available. Interpretation can occur across languages or within one language. Unlike direct quotation, indirect quotation is not a mere reproduction. Nor is, of course, translation, which consists in finding one sentence s in language *L* that is synonymous (which has the same meaning) with a distinct sentence *s*' from language *L*'. Offering a summary or an exegesis of an utterance, which may take place within one language, involve finding other sentences in one and the same language to help determine the content of the initial utterance. The contents of two or more representations can stand in a *similarity* (or resemblance) relation of which the identity relation is, at best, an extreme limiting case.[12]

What is true of the contents of linguistic representations is true *mutatis mutandis* of the contents of mental representations. Unlike linguistic representations, mental representations lack phonological properties. Arguably though, assuming, as I do, that mental representations have neurological properties, techniques of brain imaging, might, unlike ordinary mindreading, allow one to reproduce a mental representation in virtue of copying some of its neurological properties. However, this very process in and of itself does not display the meaning or content of the copied mental representation. Only by *interpretation*, not by copying, can one get at the content or meaning of any representation. Only by creatively providing another representation can one interpret the content of a representation.

According to Gordon's (1986, 1995, 1996) anti-introspectionist version of the simulation theory, to ascribe a belief to someone is to engage in *pretence*.[13] To ascribe a belief to someone else is to mimick the other's mental state by going into a pretend-state. As Gordon (1986, p. 68) puts it, "to attribute a belief to another person is to make an assertion, to state something as a fact, *within the context of practical simulation*. Acquisition of the capacity to attribute beliefs is acquisition of the capacity to make assertions in such a context." I will now argue that the simulation (or pretence) heuristic can indeed contribute to, but cannot constitute a full account of, third person mindreading. It cannot because there is a gap between engaging in pretence and ascribing beliefs.

---

[12] Cf. Jacob (1987).

[13] Gordon (1995, p. 53) explicitly links third-person mindreading to the task of actors.

Gordon's (1986, 1995) account originates from Quine's (1960) famous observation that indirect quotation and the attribution of propositional attitudes are "essentially dramatic acts" in which we "cast ourselves in unreal roles" and "we project ourselves into what, from his remark and other indications, we imagine the speaker's state of mind to have been, and then we say what, in our language, is natural and relevant for us in the state thus feigned". Building on Quine's insight, Stich (1983, pp. 84–88) argues that when I attribute a belief to somebody else (as in "Bush believes that *p*"), I do *two* things at once: first, I "perform a little skit, doing a bit of play acting" in which the "that"-clause "constitutes the script for the play". Secondly, to say that Bush believes that *p* is to say that he is in a state similar to the state which would play the same typical role if my utterance of sentence "p" had had a typical causal history.

In other words, to say that Bush believes that *p* is to say that he is in state similar to the state I would be in were I to utter "p". The point is that Stich does not present the first part of his account – the play acting or pretence – as a *full* account of belief-ascription. On Gordon's account, however, in ascribing a belief, I merely perform a simulation, which gives rise to a pretend thought or a pretend assertion. Stich's account involves two steps, not just one like Gordon's. In other words, Gordon fails to tell us what we are supposed to do with our pretend thought or our pretend assertion (see Fuller, 1995).

On Gordon's pure pretence-theoretic account, in ascribing a belief to another person, I engage in pretence, i.e., I entertain a pretend thought or make a pretend assertion: I merely perform a little skit or a bit of play-acting and nothing else. But I think that this pure pretence-theoretic view of third-person mindreading betrays a confusion between engaging in pretence and using pretence for the purpose of representing a representation (or metarepresenting a representation). To ascribe a belief to a third person is to form a thought (or representation) about the other person's thought (or representation). The pure pretence-theoretic account has things backwards: when I ascribe a belief to a third person, I do not make an assertion or entertain a thought within the context of practical simulation. Rather, if I use simulation or pretence at all, I do so within the context of something more complex, i.e., representing a representation. Thinking about someone else's thoughts does not consist in – it does not amount to – either entertaining a pretend thought or making a pretend assertion.

I may ascribe beliefs to a third person for the purpose of deriving from her information about aspects of the world which I either cannot directly perceive myself or do not understand, for the purpose of evaluating her reliability or rationality or for the purpose of arguing with her. When I do so ascribe a belief to a third person, I exercise my concept of belief. It is constitutive of beliefs that they can be misrepresentations: if a mental representation cannot be false, then it is not a belief. So, possession of the concept of belief crucially allows one to entertain complex thoughts (of the type "Bush believes that p") involving embedded representations one knows (or believes) to be false. I can e.g., form the complex thought expressed by the sentence "George W. Bush believes that witches have magical power" or "Aristotle believed that the Earth is at the center of the universe". When I ascribe such beliefs, my belief-ascription is not a pretend thought: it is not about some possible imaginary world or situation. It is about George W. Bush's beliefs or Aristotle's beliefs. Since George W. Bush and Aristotle are actual people, my belief-ascriptions describe small portions of the actual world.

To fully appreciate the gap between engaging in pretence and using pretence for the purpose of metarepresenting a representation, consider a piece of fiction, e.g., Shakespeare's *Hamlet*. It is one thing to engage in pretence for the purpose of enjoying the play and to entertain pretend thoughts about fictitious characters and think (or say that) Hamlet was a 16th century prince of Denmark who was engaged to Ophelia and killed his uncle Claudius, whom he suspected to have murdered his own father. It is another thing to think (or say that) *Hamlet*, which is the longest of all the plays written by the English Elizabethan playwright, William Shakespeare, depicts a fictitious character named "Hamlet", who was a 16th century prince of Denmark, who was engaged to Ophelia and who killed his uncle Claudius, whom he suspected to have murdered his own father. In entertaining the former pretend thought, one pretends of a fictitious character that it is not fictitious. One's pretend thought is about some possible imaginary world containing several fictitious imaginary characters. In the latter case, one's thought is about the actual world: it is about an actual play, written by an actual playwright, and which anybody can read, see and enjoy by engaging in pretence. Presumably, human children can engage in pretence and games of make-belief before they can fully deploy the concept of belief – as witness famous false belief tasks. Engaging in pretence yields immersed (practical or engaged) forms of third-person mindreading. But human adults have the ability to ascribe, understand and evaluate (for truth and falsity) others people's beliefs in a detached reflective way. This is what lies outside the scope of the simulation theory of third-person mindreading.[14]

6. MENTAL SIMULATION AND FIRST-PERSON MINDREADING

Gordon (1996) has offered an account of first-person mindreading consistent with his anti-introspectionist version of the simulation theory and which he calls the "ascent routine" (and which was anticipated by Evans 1982). To see how the "ascent routine" works, consider a question about myself such as: "Do I believe that George W. Bush is the new president of the US?". On the ascent routine model, in order to answer this question, I switch to a logically simpler question: "Is George W. Bush the new president of the US?" This is a question about George W. Bush and US politics, not about myself. According to the ascent routine, if the response to the latter question is "Yes", then the answer to the former question will be "Yes" too. If the answer to the latter question is "No", then so will be the answer to the former question. It is clear why this simple strategy will not work in the general case: at best, it can provide self-ascriptions of *beliefs*, not sefl-ascriptions of any other attitude (e.g., hope, fear or desire). If the question had been instead "Do I approve of the fact that George W. Bush is the new president of the US?" or "Do I deplore the fact that George W. Bush is the new president of the US?", clearly by using the ascent routine and considering the logically simpler question "Is George W. Bush the new president of the US?", I could not answer such questions. So the ascent routine fails to have the required scope for first-person mindreading.

I now turn to Goldman's introspectionist view of first-person mindreading. Goldman (1989, 1993, 2000) holds an introspectionist version of the simulation theory of third-person mindreading. On his view, one acquires third-person understanding of the mind of others by running a simulation using one's own resources. Thus, third-person mindreading presupposes first-person mindreading. However, Goldman (2000) thinks that simulation fails as an account of psychological self-knowledge.

Goldman's (2000) reason for thinking that mental simulation fails as an account of self-knowledge is quite simple and, I think, decisive. Let us consider the off-line simulation model of my prediction of somebody else's decision. Suppose I am trying to predict a decision you are about

---

[14] For further discussion, see Jacob (2002) and Dokic (2002).

to make and I am using the pretence-based simulation routine. On this model, I create in myself some pretend beliefs and desires that are supposed to match your beliefs and desires – those beliefs and desires on which you are basing your decision (which I am trying to predict). I feed these pretend states into my practical reasoning system; I turn it on; and it supplies a pretend-decision. Of course, I do not act on this pretend-decision. Rather, I store this pretend-decision in my belief-box as information about you and use it to predict your decision. So, on this model, using a simulation pretence-theoretic strategy to do third-person mindreading depends on or presupposes the ability to do first-person mindreading.

But now the question naturally arises: how did I detect the output of my decision-making system? How did I determine which decision was selected by my practical reasoning system on the basis of the pretend beliefs and desires which served as inputs? Why could I not use the pretence-theoretic simulation heuristic to respond to this question and treat myself in a third-personal way? The reason why I cannot is the following. Were I to use simulation to identify the output of my own decision-making system, I would be involved in an infinite regress and I would never succeed in the task. In order to identify the result of my practical reasoning system, I would have to use a second-order simulation heuristic. This second-order simulation heuristic would generate a second-order output. Now, if I need a second-order simulation heuristic to determine the output of my first-order simulation, then presumably, I could not identify the second-order output of the second-order simulation unless I used a third-order simulation heuristic, which would generate a third-order output, and so on endlessly. But presumably, I can identify the output of my first-order simulation heuristic. And so, I do not use the simulation heuristic to read my own mind.

Presumably, I can simulate myself from the perspective of someone else.[15] But on the one hand, this involves an embedded use of simulation since I cannot directly simulate myself from someone else's perspective: I can only simulate myself from the *simulated* perspective of someone else, i.e., if I simulate the other's perspective. On the other hand, I can simulate myself from the simulated perspective of someone else – for fun – while firmly believing that the other person's beliefs about me are *wrong*, not right. So the question is: how do I read my own mind? What underlies direct immediate introspective self-knowledge? Goldman has

---

[15] See Dokic (2002, p. 115).

sketched an answer to this question. According to him, we have what he calls "an inner detection mechanism" attuned to identifying our own mental states. In the past, Goldman has argued that propositional attitudes (beliefs and desires), as well as sensory states, have so-called phenomenal properties, i.e., intrinsic qualitative properties. If they do, then these phenomenal properties would presumably be the targets of the inner detection mechanism.

First of all, I want to point out that Goldman's proposal will not get off the ground unless beliefs and desires do indeed have phenomenal properties. I am skeptical. For example, I doubt very much that there is any phenomenology to believing that 5 is a prime number. I think that beliefs can give rise to emotions and emotions do have intrinsic phenomenological properties. But the idea that beliefs have intrinsic phenomenological properties might arise from confusion between beliefs and belief-caused emotions. Plainly, one and the same proposition held under the same attitude (i.e., belief) can give rise to different emotions in two different persons at the same time in virtue of the differences between their belief systems and their values. Think for example of the proposition that Bill Clinton was not impeached in the fall 1998. Presumably, both Kenneth Starr and Bill Clinton can hold the belief that this proposition is true. However, if they do, this single belief will not prompt the same emotional response in both.

Secondly, Goldman's quasi-perceptual model of first-person mindreading seems open to the objections against the traditional empiricist conception of psychological self-knowledge. First of all, if one's knowledge of psychological facts about oneself derives from some quasi-perceptual acquaintance with one's own perceptual experiences, then it is questionable whether – as empiricist philosophers (but not Goldman) have claimed – self-knowledge can still meet the epistemic requirements of genuine a priority, let alone infallibility. Secondly, as many philosophers have noticed, whereas vision, audition, olfaction, touch and proprioception can be used to get information about mind-independent facts (involving one's own body and the bodies of others), no inner sense organ provides information about one's own perceptual experiences, let alone about one's own thoughts. Finally, as Harman (1990), Tye (1992), Shoemaker (1996) and many others have noticed, perceptual experiences are introspectively transparent (in a sense of transparency different from the Cartesian sense): they are diaphanous. What one experiences are mind-independent objects, not one's own experiences. One sees, hears or smells non-mental things by having visual, auditive or olfactory experiences. But one one does not see, hear or smell one's own visual, auditory or olfactory experiences. In other words, the phenomenology of the introspection of one's own visual experience of a bush of blue lavender just is the phenomenology of one's visual experience of a bush of blue lavender. What it is like to introspect and to be aware of one's own visual experience of a bush of blue lavender is nothing other than what it is like to have the visual experience of a bush of blue lavender. Presumably, if introspection of one's visual experience of anything involved some quasi-perceptual process, then introspective awareness of one's visual experience would have a phenomenology of its own – in addition to and above that of the visual experience itself.

More generally, the quasi-perceptual model of self-knowledge seems open to Shoemaker's (1996) powerful objections. Shoemaker (1996, pp. 204–6), who accepts the distinction between perceiving objects and perceiving facts, lists eight conditions all of which are satisfied by the "object perception model" and three of which are satisfied by what he calls the "broad perceptual model." (1) Sense perception results from an organ of perception. (2) Sense perception involves a distinctive phenomenology. (3) Perceptual awareness of facts is mediated by perceptual awareness of objects. (4) Sense perception of an object leads to identification of the object. (5) Sense perception of an object involves a causal relation between a sense organ and the object's intrinsic properties. (6) Objects of perception can be tracked by attention. (7) Perceptual beliefs are caused by the facts involving the perceived objects. (8) Facts involving perceived objects are independent from the perceptual mechanisms.

If, as I think, Shoemaker's eight conditions do fix the boundaries of sense perception, then Goldman's quasi-perceptual model of introspective self-knowledge is in trouble. Whenever one knows what one experiences or what one believes (whether one's belief is occurrent or dispositional), one is not made aware of one's experience or of one's belief by means of a sensory experience of either one's experience or of one's belief. One becomes aware of either one's experience or one's belief by means of a conceptual higher-order representation – or metarepresentation – of one's experience or of one's belief. In other words, one would not be aware of one's experience or of one's belief unless one had the concept of *experience* and the concept of *belief*.[16]

---

[16] See Rosenthal (1993, this volume) for elaboration of the notion of higher-

Although I embrace Shoemaker's (1996) critique of the perceptual model of self-knowledge, I do not think that his criticism forces one to accept his further view (which I do not accept) that there is a conceptual (non causal) constitutive relation between one's first-order mental states (experiences and propositional attitudes) and self-knowledge. What Shoemaker (1996, pp. 225–7) calls "self-blindness" is, on his view, a conceptual impossibility. A creature cannot have experiences and propositional attitudes and not be aware that she has them. In other words, it is constitutive of a creature with beliefs and experiences that she cannot be "self-blind". I think that this claim is too strong. I accept Shoemaker's (1968) claim that psychological self-knowledge is "immune to error of misidentification" in the following sense. Psychological self-knowledge is not infallible, but since it is not perceptual, it does not involve a process of self-identification. Although fallible, it is, nonetheless, immune to the kind of error to which perceptual processes are exposed.

In between the perceptual model of self-knowledge and the claim that self-blindness is conceptually impossible, there is, I think, room for an intermediary view according to which self-knowledge is neither perceptual nor conceptually inevitable. One may certainly fail to achieve (at least full) psychological self-awareness. I agree with Shoemaker that one's higher-order representations (or metarepresentations) of one's experiences or beliefs are conceptual, not perceptual, representations. But this is consistent with denying that there is a constitutive link between one's experience or one's belief and one's higher-order reepresentation of one's experience or of one's belief.

REFERENCES

Baron-Cohen, S. (1995). *Mindblindness*, Cambridge, Mass.: MIT Press.

Bermudez, J. (1998). *The Paradox of Self-Consciousness*. Cambridge, Mass.: MIT Press.

Campbell, J. (1994). *Past, Space and Self*. Cambridge, Mass.: MIT Press.

Carey, S. & Spelke, E. (1994). Domain specific knowledge and conceptual change. In *Mapping the Mind, Domain Specificity in Cognition and Culture*, ed. L. Hirschfeld & S.A. Gelman. Cambridge: Cambridge University Press, pp. 169–200.

order thought for a different purpose, namely for the purpose of accounting for the difference between conscious and unconscious mental states.

Currie, G. (1995). Visual imagery as the simulation of vision. *Mind and Language* 10(1/2): 25–44.

Davidson, D. (1984). First-person authority. In Davidson, D. *Subjective, Intersubjective, Objective*. Oxford: Oxford University Press 2001, pp. 3–14.

Davidson, D. (1987). Knowing one's own mind. In Davidson, D. *Subjective, Intersubjective, Objective*. Oxford: Oxford University Press 2001, pp. 15–38.

Dokic, J. (2002). Reply to Pierre Jacob. In *Simulation and Knowledge of Action*, ed. J. Dokic & J. Proust. Benjamin, pp. 111–117.

Evans, G. (1982) *The Varieties of Reference*. Oxford: Oxford University Press.

Fuller, G. (1995). Simulation and psychological concepts. In *Mental Simulation*, ed. M. Davies & T. Stone. Oxford: Blackwell 1995, pp. 19–32.

Gallese, V. (2003). The manifold nature of interpersonal relations: the quest for a common mechanism. *Philosophical Transactions of the Royal Society of London* 358: 517–28.

Gallese, V. & Goldman, A. (1998). Mirror neurons and the simulation theory of mind-reading. *Trends in Cognitive Sciences* 2(12): 493–501.

Gibson, J.J. (1979). *The Ecological Approach to Visual Perception*. Boston: Houghton-Miffin.

Goldman, A. (1989). Interpretation psychologized. In *Folk Psychology*, ed. M. Davies & T. Stone. Oxford: Blackwell 1995, pp. 74–99.

Goldman, A. (1993). The study of folk psychology. *Behavioral and Brain Sciences* 16: 15–28.

Goldman, A. (2000). The mentalizing folk. In *Metarepresentations: A Multidisciplinary Perspective*, ed. D. Sperber. Oxford: Oxford University Press, pp. 171–196.

Gopnik, A. & Meltzoff, A. (1997). *Words, Thoughts and Theories*. Cambridge, Mass.: MIT Press.

Gopnik, A. & Wellman, H.M. (1992). Why the child's theory of mind really is a theory. In *Mental Simulation*, ed. M. Davies & T. Stone. Oxford: Blackwell 1995, pp. 232–258.

Gopnik, A. & Wellman, H.M. (1994). The theory theory. In *Mapping the Mind, Domain Specificity in Cognition and Culture*, ed. L. Hirschfeld & S.A. Gelman. Cambridge: Cambridge University Press, pp. 257–293.

Gordon, R. (1986). Folk psychology as simulation. In *Folk Psychology*, ed. M. Davies & T. Stone. Oxford: Blackwell 1995, pp. 60–73.

Gordon, R. (1995). Simulation without introspection or inference from me to you. In *Mental Simulation*, ed. M. Davies & T. Stone. Oxford: Blackwell 1995, pp. 53–67.

Gordon, R. (1996). 'Radical' simulationism. In *Theories of Theories of Mind*, ed. P. Carruthers & P.K. Smith. Cambridge: Cambridge University Press, pp. 11–21.

Harman, G. (1990). The intrinsic quality of experience. In *The Nature of Consciousness*, ed. N. Block, O. Flanagan & G. Güzeldere. Cambridge, Mass.: MIT Press 1997, pp. 663–675.

Hauser, M.D., Chomsky, N. & Fitch, W. T. (2002). The faculty of language, what is it, who has it, and how did it evolve. *Science* 298: 1569–79.

Jacob, P. (1987). Thoughts and belief-ascriptions. *Mind and Language* 2(4): 301–25.

Jacob, P. (2002). The scope and limits of mental simulation. In *Simulation and Knowledge of Action*, ed. J. Dokic & J. Proust. Benjamin, pp. 87–109.

Jacob, P. & Jeannerod, M. (2003). *Ways of Seeing, the Scope and Limits of Visual Cognition*. Oxford: Oxford University Press.

Jeannerod, M. (1997). *The Cognitive Neuroscience of Action*. Oxford: Blackwell.

Jeannerod, M. (2001). Neural simulation of action: a unifying mechanism for motor cognition. *Neuroimage* 14: S103–S109.

Jeannerod, M. (2002). The mechanism of self-recognition in humans. *Behavioural Brain Research* 142: 1–15.

Leslie, A.M. (1994). Pretending and believing: issues in the theory of ToMM. *Cognition* 50: 211–38.

Leslie, A.M. & Thaiss, L. (1992). Domain specificity in conceptual development: neuropsychological evidence from autism. *Cognition* 43: 225–51.

Meltzoff, A.N. (1995). Understanding intentions of others: reenactment of intended acts by 18 months-old children. *Developmental Psychology* 31: 838–50.

Meltzoff, A.N. & Moore, M.K. (1977). Imitation of facial and manual gestures by human neonates. *Science* 198: 75–78.

Meltzoff, A.N. & Moore, M.K. (1997). Explaining facial imitation. A theoretical model. *Early Development and Parenting* 6: 179–192.

Perrett, D.I., Rolls, E.T. & Caan, W. (1982). Visual neurones responsive to faces in the monkey temporal cortex. *Experimental Brain Research* 47: 329–42.

Perrett, D.I., Harries, M.H., Bevan, R., Thomas, S., Benson, P.J., Mistlin, A.J., Chitty, A.J., Hietanen, J.K. & Ortega, J.E. (1989). Frameworks of analysis for the neural representation of animate objects and actions. *Journal of Experimental Biology* 146: 87–113.

Quine, W.V.O. (1960). *Word and Object*. Cambridge, Mass.: MIT Press.

Rizzolatti, G., Camarda, R., Fogassi, L., Gentilucci, M., Luppino, G. & Matelli, M. (1988). Functional organization of area 6 in the macaque monkey. II. Area F5 and the control of distal movements. *Experimental Brain Research* 71: 491–507.

Rizzolatti, G., Fadiga, L., Gallese, V. & Fogassi, L. (1995). Premotor cortex and the recognition of motor actions. *Cognitive Brain Research* 3: 131–141.

Rizzolatti, G., Fogassi, L. & Gallese, V. (2000). Cortical mechanisms subserving object grasping and action recognition: a new view on the cortical motor functions. In *The New Cognitive Neurosciences*, ed. M.S. Gazzaniga. Cambridge, Mass.: MIT Press, pp. 539–552.

Rosenthal, D.M. (1993). Thinking that one thinks. In *Consciousness: Psychological and Philosophical Essays,* ed. M. Davies & G.W. Humphreys. Oxford: Blackwell, pp. 197–223.

Rosenthal, D.M. (this volume). Consciousness, Interpretation, and Higher-order thought.

Shoemaker, S. (1968). Self-knowledge and self-awareness. *The Journal of Philosophy* 65(19): 555–67.

Shoemaker, S. (1996) *The First-Person Perspective and Other Essays*. Cambridge: Cambridge University Press.

Spelke, E.S. (2000). Core knowledge. *American Psychologist* 55: 1233–1243.

Stich, S. (1983). *From Folk Psychology to Cognitive Science*. Cambridge, Mass.: MIT Press.

Stich, S. & Nichols, S. (1992). Simulation or Tacit Theory? In *Folk Psychology*, ed. M. Davies & T. Stone. Oxford: Blackwell 1995, pp. 123–158.

Stich, S. & Nichols, S. (1995) Second Thoughts on Simulation. In *Mental Simulation*, ed. M. Davies & T. Stone. Oxford: Blackwell, pp. 87–108.

Stroud, B. (1989). Understanding human knowledge in general. In *Knowledge, Readings in Contemporary Epistemology,* ed. S. Bernecker & F. Dretske. Oxford: Oxford University Press 2000, pp. 307–323.

Tomasello, M. (2000). *The Cultural Origins of Human Cognition*. Cambridge, Mass.: Harvard University Press.

Tye, M. (1992). Visual qualia and visual content. In *The Contents of Experience*, ed. T. Crane. Cambridge: Cambridge University Press, pp. 158–176.

B. Current Interdisciplinary Research: Neuroscience and Psychoanalysis

VII. FREUD'S PROJECT FOR A SCIENTIFIC PSYCHOLOGY IN THE 21ST CENTURY

KARL PRIBRAM

The recent resurgence of interest in biology that has invigorated psychiatry has penetrated the hallowed halls of psychoanalysis. Now, therefore more than ever, Freud's "Project for a scientific psychology" (Freud 1950a [1895]) has relevance for those working within the general framework provided by the psychoanalytic metapsychology. In this essay I shall summarize some of Freud's propositions regarding brain function in the light of what we have learned during the past century.

My interest in "The Project" was aroused by two propositions:

1. a memory based theory of motivation and
2. an Ohm's law of neural processing.

Two other interests developed as Merton Gill and I began our studies prior to the publication of our book *Freud's Project Reassessed* (Pribram & Gill 1976):

3. the brain mechanisms involved in primary and secondary processes and
4. brain mechanisms in unconscious and especially conscious processes.

1. THE MEMORY BASED THEORY OF MOTIVATION

During the 20th century drive theories of motivation flourished. Freud began this trend with his (or his patients') discovery of the untrustworthiness of remembrances of childhood traumatic experiences. Many arguments have recently occurred regarding Freud's honesty in retreating from a veridical memory interpretation and more generally the legal status of memory as evidence. My stance is that there is record that Freud was deeply upset by the discovery that necessitated a complete revision of his theoretical stance. Accepting the importance of drive stimuli in distorting memory did not come easily. And Freud had little to lose or gain by the revision: he was already castigated as a rebel because of his emphasis on sexuality, adding that libido could distort remembering did not add (or subtract) much from how he was then viewed.

But the discovery of the fallibility of memory did shake the very foundation upon which Freud's talking cure was based. The Wuerzbergers had shown that once a question had been thoroughly understood and accepted, the answer was preordained irrespective of the paths of thought taken to reach the conclusion. Franz Brentano (as did William James) highlighted the issue by noting that now the ACT of Thinking needed to be investigated. Freud, having taken 7 courses from Brentano came up with the technique of allowing "free associations" to be vented. (The freedom was of course automatically constrained by the patient's problem that needed solution). Freud stated that his hope was that the cognitive structure of the memory process would by this technique be uncovered: voila, the issue of "The Act of Thinking" was addressed. But if remembering could be skewed by current drive stimuli the cognitive structure became as mysterious as the dancer covered by her seven veils.

In today's purview one would simply hold that memory is forged by both sensory and biochemical receptor input, and being reconstructive, is vulnerable to current input from both sources. How much distortion occurs needs to be painstakingly established in each specific situation.

2. OHM'S LAW OF NEURAL FUNCTION

Freud has often been misunderstood as holding a hydraulic view of the neural process. This is in part due to the use of this metaphor in discussing electrical circuitry and in part due to translations of "The Project" by persons who did not know the neurological terminology in use when Freud was writing and even now. Thus "action currents" become currents in flow and local potential changes (Besetzte Energie, QsubN) become cathexes. These quantities of local energy accumulate because of resistances at contact barriers (synapses) and act much as capacitors to facilitate discharge and thus the generation of action currents.

What intrigued me was the inclusion of local potential changes in the process, a much neglected inclusion even today when neural circuitry based on axonic propagation of spikes (action currents) is invoked to the exclusion of telendrendronic pre- and dendritic post-synaptic local processing. It is in the local oscillatory hyper- and de-polarizations that much of the processing that is coordinate with thinking occurs. As Sherington (1906) as well as Freud noted, the more automatic a process "the less that mind accompanies it". More on this under the topic of consciousness.

3. PRIMARY AND SECONDARY PROCESSES

Freud made a distinction between primary and secondary processes, a distinction that is missed despite the current popularity of cognitive science and cognitive neuroscience. Primary processes are associative. Secondary processing is hierarchical consisting of a double loop of attentional matching involving reality testing. Much current thinking in cognitive science is still constrained (shackled in fact) by an attempt to fit processing into an assocaiative frame. Attention is conceived as being allocated within a limited fixed resource. Closer to Freud's view is the fact that the resource is flexible and sensitive to both the situation in which It occurs and the organism's current state of readiness (see Pribram & McGuinness 1975, 1992). Effort (reality testing) is expended when the situation and the readiness state are not readily reconciled.

For Freud cognition itself is much more akin to what we now conceive as a computer program made up of lists that are addressed. Items in the list contain the code that instructs the next step to be taken in action or thought. The code itself may be associative, that is, primary process but it is utilized in the service of a hierarchically arranged cognitive structure (see Pribram & Gill 1976, pp. 72–75 for evidence).

Freud was a cognitivist. He coined the term agnosia, wrote a book *On Aphasia* [Freud 1891b] which is still used. One might even say that the current practice of experimental and clinical psychology and of psychiatry is much more in tune with the practices of the late 19th century than with those of the 20th. This is especially true of current views of the role of conscious experience in our dealings with the world we inhabit.

4. BRAIN AND CONSCIOUS EXPERIENCE

Freud is best known for his emphasis on unconscious processes. As did all neuroscientists in Vienna at the time he attributed the organization of conscious experience to cortical processing. He was intrigued by what the rest of the brain might be doing. I was intrigued by the processes he attributed to the cortex: that cortical function is based on "patterns of periodicity" transmitted from the senses. By the time Merton Gill and I began to seriously study "The Project" I had embraced the idea that some important cortical functions are holographic-like. Holograms encode the transformation of space-time patterns into spectra that are composed of interferences among waveforms measured in terms of their frequency. Frequency is the inverse of periodicity and Freud's idea regarding patterns of periodicity thus comes close to the idea that cortical processing operates (in part) in terms of interference patterns. My hypothesis has been that dendritic receptive fields constitute such patterns and that conscious experience is based on what is going on in these fields (Pribram 1966, 1971, 1991).

I noted earlier that Freud had envisioned a double loop mechanism of attention that entailed reality testing. Following Sherington I proposed that a "temporal hold" has to operate to delay processing for awareness to be experienced. Recent evidence has shown that such a delay is imposed by the frontal cortex on the systems of the posterior convexity (reviewed in Pribram 1999). For example a visual input activates the occipital cortex, then the frontal cortex, and then again the occipital cortex.

This delay has consequences for how experience becomes conscious. Philosophers have argued as to whether consciousness supervenes on brain processes or whether consciousness is an epiphenomenon. Neuroscientists have been puzzled by Ben Libet's findings (1966) that stimulation of the somato-sensory-motor cortex is not "sensed" for a quarter to a half second after the onset of the stimulation (whereas peripheral stimulation is sensed immediately). An answer may be ventured for the neuroscience finding: peripheral stimulations engage a much larger cortical field than do Libbet's cortical stimulations. Electrical excitation of the sciatic nerve, for instance, evokes reposes over the entire central part of the cerebral convexity (including the so-called motor and premotor cortex) even in anesthetized monkeys (Malis, Pribram & Kruger 1953). In addition, local vascular changes may be relayed to the mediobasal (limbic) motor cortex (Kaada, Pribram & Epstein 1949; Pribram 1961).

The classical precentral "motor" cortex is, in fact a sensory cortex for action (the evidence for this statement is reviewed in Pribram 1971, 1991). As such it encodes the environmental consequences of an action, not just the movements necessary to carry out the act. When the precentral process contributes to awareness it is of the errors in the consequences not the trajectory of the movements by which to accomplish the action. The parallel in vision is that we do not sense saccads, only the visual image effected by them. That image and the environmental consequences of an action come to awareness some time later than the movement itself.

Stimulation of the classical central (Rolandic) sensory and motor cortex should not be coordinate with awareness. If we were aware of our actions at the time they are occurring, we would mess them up (Miller, Galanter & Pribram 1960; Pribram 2003). Imagine being aware of your tongue and palate as you are giving a talk – in fact occasionally when your mouth becomes dry, you do become aware and just can't go on. Or playing tennis or batting at baseball – the adage is "keep your eye on the ball". When taking notes during a lecture, conscious attention is on what the lecturer is saying not the writing of notes. Furthermore, the muscle contractions, the movements involved can vary according to whether one is using a writing pad, a laptop computer or standing at the blackboard. The primary sensory and motor systems provide the encoded intended consequences of an action, not just the particular movements needed to carry them out (see Pribram 1971, 1991 for review of the evidence for these statements). Thus, these systems need to function autonomously during the course of an action; only when, after a temporal hold, they act in concert with other brain systems do they participate in organizing any necessary change in future acts by way of conscious intervention.

There is another piece of evidence that supports Freud's double loop model of attention. When we first began to study event related brain electrical potential changes (ERPs) we learned a great deal by using what is called an odd-ball technique. A particular stimulus is presented repeatedly and a different (but somewhat related) stimulus is randomly interposed in the series. The recorded ERPs are then averaged separately for the two types of stimuli. The averaged records are dissimilar especially around 300 millisec. after stimulus presentation. We interpreted the change in the ERP for the odd-ball stimulus as indicating that an update in the perception of the stimulus sequence was occurring. But subsequent experimentation showed that another dissimilarity in ERPs could be observed at around 400 millisec. and that updating did not occur unless the 400msec. dissimilarity was present. In short, the dissimilarity at 300 millisec. indicated that an update was necessary and the 400 miliisec. heralded the actual updating. According to all this evidence (Pribram & McGuinness 1992) attention is a two stage process.

At one point William James declares that he is tired of all the confusion surrounding consciousness; that we should stick to trying to understand attention. But of course neither he nor Freud did so. I believe there are several critical elements that contribute to the confusion. One of these is conflating of sensory processing and control of movement with conscious experience. Second, two levels of conscious experience fail to be distinguished: Freud called them preconscious and conscious; today we use such terms as perception and reflective consciousness (Pribram 1999). These conflations have led philosophers to opt for one of two explanations. One is that all experience directly supervenes on brain processes and the other is that consciousness is an epiphenomenon and irrelevant to living adaptively in the world. I believe both are in error.

The two stages involved in attention provide the key to an alternative explanation. Experience does not immediately supervene on neural processing during a perception or an action. Both are processes that, at any moment, are more like feedforward programs than error sensitive to feedback. We are not aware of the process by which we prehend an object. As noted, this is a good thing – we'd only mess-up. So, does that leave us with conscious experience as an epiphenomenon? Not at all. After I reflexly remove my hand from a hot flame, I contemplate the happening. Our cat is an excellent example: he looks at his paw and licks it – then looks at the offending object and reaches out toward it but this time does not touch it. He repeats this procedure several times over. If I may anthropomorphize, the cat's conscious awareness of the incident, his ERP at 300 millisec. and the later 400 millisec. indicate that awareness of what has happened alter future behavior. In the example given, the cat reinforced the change in subsequent behavior several times – in operant conditioning terms he was shaping the changes in his behavior. And the non-behavioristic claim is that the shaping occurred by way of conscious awareness of what was happening. Conscious experience is not an epiphenomenon. Conscious experience shapes subsequent behavior.

Nor does experience immediately supervene on ongoing neural processing. Consciousness of an experience when attained affects subsequent brain processes by changing the consequent contingencies within which the brain processes now occur. How then does a pattern of thought that characterizes a conscious experience influence an ongoing processing pattern in the nervous system?

Are the patterns that we experience identical with those that characterize brain function? I don't believe so. My view takes computer programming as a metaphor. I address the hardware my computer by using a natural language such as English. But the computer uses BIT language. To communicate an operating system intervenes: A binary code is used to address the hardware. This code is converted into an octal or hexadecimal, the hexadecimal into an alphabetical language using few symbols to make up short words and this language in turn is transformed into higher and higher order languages such as ASKII, ALGOL, JAVA and so forth. At the top of the hierarchy is the word processing language that can communicate with me. For the output the reverse sequence is put into operation. Each level of processing language converts the prior one which uses fewer symbols over a longer series to a shorter series using a greater range of symbols. A Cartesian dualist, knowing nothing about operating systems might well conclude that natural languages and hardware codes (mind talk and brain talk) have nothing in common.

I have repeatedly reviewed the evidence (Pribram 1991, 1999) for stating that Gabor functions (Gaussian constrained sinusoids) or some similar four dimensional hyperspaces based on Jacoby delta functions or Wigner distributions can serve the mind/brain connection the way the BIT language serves present day computers. Gabor invented his wavelet to determine the maximum compression that one could place on a telephone communication across the Atlantic Cable. (It came out to be approximately half a cycle for each wavelength). Gabor used the same mathematics that Heisenberg had used to characterize subatomic units in physics. Gabor therefore called his unit of communication a quantum of information. This unit shares with Heisenberg's the indeterminacy on measurement at the lower limit. Gabor related this minimum uncertainty in measurement to Shannon's reduction of uncertainty in the bit language domain.

The Gabor function has been found to accurately describe the dendritc receptive fields of sensory neurons in the brain cortex. The Gabor quantum of information can therefore serve the same function for the wetware/experience relationship that the BIT serves for the hardware/software relation.

The metaphor goes further in that just as hierarchically arranged structures compose the current day hardware so hierarchical structures compose the brain modules and systems. And in a manner similar to programming languages our experience becomes coded hierarchically (as for instance, unconscious, preconscious, and self conscious). But just as the hierarchical hardware structure of the computer has little semblance to the hierarchical structure of programming languages, so the hierarchical structure of brain processing is not that of the hierarchical structure of our experience. There are no pictures in the brain, only quantum holographic (holonomic) processes based on Gabor-like wavelets. To use another metaphor, the processing of an fMRI tomograph uses quantum holography, the pictures we see are reconstructions made possible by the process. In a similar fashion, as already noted, our actions are controlled by brain processes that, only at the lowest level of a hierarchy, match the patterns of movements and muscle contraction (see Llinas 2001).

A final word: What about the patterns that characterize our thought processes? Do they supervene directly onto patterns of brain activity? Freud as well as many others defined thought as implicit action and based his talk therapy on that principle. According to the view that I have here assembled, implicit action when we become aware of it, that is, when we consciously think about something, we involve the body's effectors, muscles and glands. Watson was not far off in his physiological behaviorism. Evidence continues to accumulate that very slight changes in muscle tone or in breathing or heart rate variability occur during thinking. William James and more recently Antonio Damasio have called attention to the involvement of feelings as bodily responses to what happens and how these feelings influence cognition and decision making (Damasio 1999). What I am making explicit is that the outcome of these body responses to conscious experience is what changes brain patterns so that subsequent experience and intended action become modified.

Conscious experience is not an epiphenomenon nor do its patterns directly supervene onto the patterns of brain processes. Conscious experience must be "taken to heart" and acted upon, albeit even minutely and tentatively, to change the brain process.

Thus Freud in the 21st Century.

REFERENCES

Damasio, A.R. (1999). *The Feeling of What Happens.* New York: Harcourt Brace & Co.

[Freud, S. (1891b). *On Aphasia.* Partly in *S.E* 14: 206–215.]

Freud, S. (1950a [1895]). Project for a scientific psychology. *S.E.* 1: 295–387.

Kaada, B.R., Pribram, K.H., & Epstein, J.A. (1949). Respiratory and vascular responses in monkeys from temporal pole, insula, orbital surface, and cingulate gyrus. *Journal of Neurophysiology* 12: 347–356.

Libet, B. (1966). Brain stimulation and conscious experience. In *Brain and Conscious Experience*, ed. J.C. Eccles. New York: Springer, pp. 165–161.

Llinas, R. (2001). *I of the Vortex. From Neurons to Self.* Cambridge, Mass.: MIT Press.

Malis, L.I., Pribram, K..H. & Kruger, L. (1953). Action potentials in "motor" cortex evoked by peripheral stimulation. *Journal of Neurophysiology* 16: 161–167.

Miller, G. A., Galanter, E.H., & Pribram, K.H. (1960). *Plans and the Structure of Behavior*. New York: Holt, Rhinehart and Winston.

Pribram, K.H. (1961). Limbic System. In *Electrical Stimulation of the Brain*, ed. D.E. Sheer. Austin: University of Texas Press, pp. 311–320.

Pribram, K.H. (1962). The neuropsychology of Sigmund Freud. In *Experimental Foundations of Clinical Psychology*, ed. A.J. Bachrach. New York: Basic Books, pp. 442–468.

Pribram, K.H. (1966). Some dimensions of remembering: steps toward a neuropsychological model of memory. In *Macromolecules and Behavior*, ed. J. Gaito. New York: Academic Press, pp. 165–187.

Pribram, K.H. (1971). *Languages of the Brain*. New York: Lawrence Erlbaum.

Pribram, K.H. (1991). *Brain and Perception. Holonomy and Structure in Figural Processing*. Hillsdale: Lawrence Erlbaum.

Pribram, K.H. (1994). What is mind that the brain may order it? In *Proceedings of Symposia in Applied Mathematics. Proceedings of the Norbert Wiener Centenary Congress 1994*. Vol. 2, ed. V. Mandrekar & P.R. Masani. Providence: The American Mathematical Society, pp. 301–329; reprinted (1997). *The Noetic Journal* 1: 2–5.

Pribram, K.H. (1999). Brain and the composition of conscious experience. *Journal of Consciousness Studies* 6: 18–42.

Pribram, K.H. (2004). Consciousness Revisited, in preparation.

Pribram, K.H. & Gill M. (1976). *Freud's Project Re-assessed*. New York: Basic Books.

Pribram, K.H. & McGuinness, D. (1975). Arousal, Activation and Effort in the control of attention. *Psychological Reviews* 82: 116–149.

Pribram, K.H. & McGuinness, D. (1992). Attention and preattentional processing: Event related brain potentials as tests of a model. *Annals of the New York Academy of Sciences* 658: 65–92.

Sherrington, C. (1906). *The Integrative Action of the Nervous System*. New Haven: Yale University Press 1947.

VIII. EMOTION AND THE ROLE OF THE BODY IN MENTAL LIFE

REGINA PALLY

> "[By] means of my psychical work, I have had to overcome a psychical force in the patients which was opposed to the pathogenic ideas becoming conscious [...] They were all of a distressing nature calculated to arouse the affects of shame, of self-reproach and of psychical pain and the feeling of being harmed; they were all of a kind that one would prefer not to have experienced, [...] that one would rather forget."
>
> (Freud 1895d, pp. 268–269)

1. INTRODUCTION

Since its inception, psychoanalysis recognized the close connection between mind and body, *psyche* and *soma* (Freud 1950a [1895]). Although Freud believed in a biologic basis for psychological phenomena (Freud 1950a [1895]) he was unable to demonstrate it, given the limited scientific knowledge of his time. Pyschoanalysis therefore welcomed the decade of the brain, with its rich discoveries, many, of which support and enhance psychoanalytic ideas. Using current neuroscience data, I intend to show that the mind and body are linked through the activation of emotion.

2. THE NEUROSCIENCE MODEL OF EMOTION

Embedded within emotion itself is the organization of adaptive behavior (Damasio 1995; LeDoux 1994, 1996). More specifically, emotion organizes the behaviors the animal will need to adapt to its environment and organizes all the necessary brain and body changes to enact that behavior. In emotion, mind and body[1] join forces to generate and sup-

[1] The body includes the brain as well as the body proper.

port adaptive behaviors. Body, refers to the body proper with all the internal viscera, including the brain.

Neuroscience agrees with the psychoanalytic view that a behavioral response, which may have been adaptive in one situation, may prove *mal*adaptive in another. For example, one woman patient, a business executive, who as a child had an over controlling parent, learned to cope by being overly compliant to avoid conflict. However, later on, in the workplace, her over-compliance inhibits her success in a competitive career. Another area of compatibility is that both neuroscience and psychoanalysis recognize that the link between an emotion and its ensuing behavior is often outside of conscious awareness. A person can be afraid and run, or can be ashamed and lower their head and fall silent, without even being at all conscious of their feeling. By becoming more consciously aware of the link, the person has a greater ability to shift to a more adaptive behavior. For example one patient, a tenured professor at a major university, who was not aware of his deep need for admiration, casually hands me his curriculum vitae, and says "I thought this might be of interest to you" and quickly proceeds to talk about some problem with his wife. When I comment on the CV, he falls silent. As we explore what happened, he tells me how mortified he now feels, since he had been completely unaware that the CV was his way of seeking out my admiration for his accomplishments. He had repressed this longing as a result of how foolish his father made him feel for showing off.

Embedded in the neuroscience model of emotion are three interrelated assumptions, that I will emphasize throughout the paper. 1. For optimal health emotion must be regulated, 2. Most emotion occurs without conscious awareness 3. Being conscious of emotion helps regulate emotion.

For optimal health emotion must be regulated. Emotion follows a natural progression from *reaction to recovery*. Emotion is regulated because after an emotion is activated, brain and body feedback systems automatically activate the mechanisms by which we recover from that emotion and return to baseline (Gunnar 1992; Degangi et al 1991).

Feedback systems are designed so that emotion occurs within an optimal range – not too much and not too little. Emotion must be regulated so that we can cope with the environment yet maintain sufficient homeostasis to support health (Davidson, Jackson et al. 2000; Davidson, Marshall et al. 2000; Silk 1998, 2000). To react emotionally is not only automatic, and involuntary it is necessary and adaptive adaptive. Pathology occurs when patients are not able, to return to a baseline state of calm following emotional arousal. One woman patient, the business executive, when she suffers a disappoint in a relationship or career goal becomes so mired in disappointment, she loses all enjoyment in her usual activites. Another male patient, a stock broker, reacts with rage to a personal slight by his wife. He withdraws and sulks, to punish her for days.

Most emotion occurs without conscious awareness. We only become conscious of a small percentage of emotion, which neuroscientists refer to as 'feelings'– sadness, joy, fear and the like (Damasio 1994, 1999). Some individuals only consciously process the *body aspect* of emotion. (Taylor 1992) These individuals may exhibit alexithymia and be susceptible to psychosomatic and somatizing disorders.

Being conscious of emotion helps regulate emotion. Although most of the time emotion is non-conscious, consciousness of feelings can facilitate voluntary control over how the emotion is expressed – for example, in the case of the stock broker, not acting on a behavior associated with a feeling, such as not sulking he feels slighted.

3. AN EVOLUTIONARY PERSPECTIVE

It is not known exactly when emotion first evolved, but we do know that it begins to flourish with the emergence of mammals and the expansion of the limbic system (LeDoux 1996). Humans share much of the same emotional circuitry with other mammals, which is why neuroscientists believe animal emotion is a good research model for human emotion.

Without emotion, lower animals are limited to a few reflexive responses to stimuli. Emotion enables mammals to respond more flexiblly to the same stimulus in a number of ways, depending on the whole context. This reaches its highest degree in humans. Touch the lowly Aplysia snail, for example, and all it can do is reflexively withdraw its gill and mantle, no matter WHO touches it, or HOW. Touch a human, and depending on which emotion is genereated, the person can run away, fight, hug, laugh, or make love...depending on WHO the touch is from (a stranger, a friend, a lover) and HOW the touch is delivered (rough, gentle, ticklish). The large number of possibilities is what enables humans to have such complex and varied social and cultural relationships.

The most *primitive emotions* are essentially positive and negative hedonic tone, and the behaviors that ensue are essentially approach/avoid reactions. In other words – if it is good, approach it; if bad, avoid it. This is essentially what the psychoanalytic concept of pleasure/unpleasure is derived from at the biological level. Babies are born with this very primitive emotional plan. A number of other emotions emerge early in the first year of life, such as fear, sadness, surprise and joy. The 'social' emotions, such as guilt, embarrassment, and pride develop by about 2–3 years. As language develops emotions become more complex. Examples are distinctions such as pride in ones country, verses pride in ones children, or sadness over losing a friend, verses sadness over losing a parent.

Emotion can be either inborn or learned. According to Ekman and Panksepp (Ekman 1993; Panksepp 1998) there are several innate, inborn 'basic' emotions. For Ekman they are anger, fear, sadness, disgust, surprise and joy. For Panksepp they are rage, fear, panic, lust, seeking, play and care. Each emotion is presumed to have its own brain circuitry and physiology. Complex emotions are learned from experience and probably do not have a specific circuitry.

4. THE FUNCTION OF EMOTION

Emotion functions to organize the mind and body toward a common goal or purpose, by coordinating the mind and body to support the behaviors which best achieve the animal's goal (LeDoux 1996; Panksepp 1998). The goal may be to find food, or to mate, or explore or find a shady spot to relax in. With an emotion all parts of the mind and body are playing on the same 'team', meaning that perception, memory, thoughts, decisions, behavior and body physiology are coordinated to help the animal meet its goal. In a state of FEAR, the MIND and BODY coordinate. The mind is vigilant, on alert for danger, more likely to perceive danger, have frightening thoughts, and remember scary events. The body is able to run away because there is an increase in heart rate, blood pressure, respiratory rate, tone of the muscles and blood flow to the muscles.

In the clinical setting, emotion organizes the transference relationship. A particular emotion coordinates the patients perceptions, memories, associations, behaviors and visceral response. For example, the emotion of anger over not being taken seriously is triggered in the transference. The patient's glaring facial expression, tense body posture and harsh tone of voice signal anger. Their memories and associations will include incidents with this same emotion. They will perceive the situation between the two of you as one in which you do not take them seriously. Shame in the transference will organize a different set of responses. Love organizes yet another set of mind and body responses.

5. EMOTION MUST BE REGULATED

Specific neurobiologic systems are in place, both for the reaction phase of emotion and for the recovery phase. Once an emotion is generated and expressed, it naturally dissipates on its own. It is often very helpful, simply to remind patients that feelings tend to fade on their own. This reduces anxiety and distress for those who believe their bad feeling will last forever.

Dysregulated states of emotion can occur. Either not enough emotion is generated, or too much is generated or recovery fails to occur. Many of the clinical situations we treat are the result of dysregulated emotion. Dysregulated emotion effects the body as well as the mind. Dysregulation of emotion can be due to inborn traits, insecure attachment, physical or sexual trauma, and neglect or prolonged separation (Silk 2000). Infants have some inborn capacity for the self-regulation of emotion, but are highly dependent on caretakers in this way (Beebe & Lachmann 1988; Stern 1985; Beebe & Lachmann 1998). Eventually, as they mature, they develop more and more capacity for regulating their own emotion. However, all humans remain dependent, to some degree, on others for affect regulation.

6. NEUROANATOMY AND NEUROBIOLOGY OF EMOTION

The brain regions involved with emotion are the limbic system, hypothalamus, [HYPO] and the brain stem, [BS] (LeDoux 1996; Damasio 1999; Schore 1994; Panksepp 1998). The limbic system includes the amygdala, [AMG], anterior cigulate, [AC]), and a "limibic extension" into the cortex, the orbitofrontal cortex [OFC]. These brain centers stimulate the production of neurotransmitters and hormones, which run the circuitry of the brain and the body's physiologic processes.

Stimulus inputs come into the brain's sensory cortex and are funneled to the AMG, and the OFC.[2] The AMG and OFC appraise those

[2] Appraisals are for the most part non-conscious.

stimuli as to their relevance to the individual. Once relevance is determined, an emotional response is generated by the AMG and OFC via signals to the HYPO and BS. The *response output* consists body physiologic changes and behaviors. Stimulus inputs come into the brain from *external* sources (i.e. the world), through the five senses (i.e. sight, sound, taste, touch and smell) as well as from *internal* sources, (i.e. 'somato-sensory' sources, including heart rate, respiratory rate, gastric motility, and blood pressure). Outputs go back, outwardly to the *world* and, inwardly to the *body* in the form of motor behavior (such as waving, smiling, talking, running), and changes of body physiology (such as blushing, sweating, crying, sighing, muscle tension). What these details highlight is the cyclic nature of the system. Response outputs themselves, can become new stimulus inputs. For example we sense our own flush of embarrassment and feel more embarrassed.

Emotion occurs every time we perceive a situation, whether or not that situation is the simple sound of a bell, or the sound one hears while listening to a Beethoven sonata. Let me emphasize – ALL experience activates emotion which is generated automatically and involuntarily, and the vast majority of the time with out conscious awareness of the emotion. Again let me emphasize – We have very little voluntary control over what emotion is generated and what body changes occur. We do however have some voluntary control over, certain aspects of the behaviors which arise. In treatment this is valuable to explain to patients. Many patients feel either guilty or ashamed of the feelings they have, as if somehow they are accountable or responsible for how they feel. Neuroscience suggests otherwise. All we can be held responsible for is HOW we behave in the presence of those feelings. In many respects this is what Freud set out to do with sexual desire and aggression. Sexual and aggressive desires are unavoidable, normal human drives. If there is excessive guilt or anxiety, they may be too heavily defended against and will be expressed in *mal*adaptive behaviors. Psychoanalysis aims to modify guilt and anxiety which leads to more adaptive sublimated behaviors.

6. 1. The first phase of emotion: Appraisal of stimuli

Each of the five senses consists of a specialized sensory end organ (i.e., retina, cochlea) which detects sensations from the external world and sends this information, via the thalamus, to the AMG, sensory cortex and OFC. In the 'somato-sensory system, specialized sensors in the internal organs send signals about the state of the body[3], via the thalamus, to the AMG, sensory cortex and PFC. Appraisals are made about these sensations with respect to their relevance to the organism. Very simple appraisals are made at the level of the AMG[4]. These can be thought of as rough assessments of the situation. Is the situation 'good vs bad?, pleasurable vs unpleasurable?, safe vs dangerous?, rewarding vs frustrating. More complex appraisals are made at the level of the OFC[5], which can be more subtle, not so black and white. An example of a complex appraisal would be 'The cake tastes good, but I need to lose weight so it is bad for me now.' Or 'It's a beautiful day for a walk in the woods, but I probably should put that off until after I finish studying.' It must be emphasized however, even though I have written these appraisals out, as if the person is thinking out loud, appraisals are generally not conscious. A child with a depressed mother can learn what facial cues on mother's face signal 'stay away now! I won't respond to you emotionally' vs 'I am OK and available to you emotionally', without ever being consciously aware of such an assessment. Similarly a patient who reads into the analyst's body language a disapproval can become quiet, without awareness of the reason for the silence. Neuroscience therefore alters somewhat the psychoanalytic view of resistence. Sometimes patients do not know why they behave or react a certain way *not* because of resistence, but because there are many brain processes, such as appraisal, which often are just not conscious.

Appraisals can be innate or learned. Innate appraisals are built into the brain as a result of genetic programming. A newborn baby innately gives more value to faces than non-faces. A young child is afraid when a balloon pops, even if they have never been exposed to any dangerous loud noises before. But most appraisals are learned. We have to learn that running into the street is dangerous or that touching a hot stove is dangerous. Generally speaking the AMG can handle simple learned appraisal. When, however, there are a number of possible appraisals to choose from, it is the OFC that is required to select the most adaptive for the current circumstances. A situation in which the mother's mouth is

[3] The state of the body involves physiologic sensations, vibration, body position, balance, muscle tension, and chemical levels (i.e., oxygen, CO_2, pH and glucose)

[4] AMG: a subcortical structure, deep within the medial temporal lobe, near the hippocampus.

[5] OFC: essentially a 'limbic' extension into the frontal lobe and part of the prefrontal cortex, which is the seat of higher cognition and executive control functions.

smiling, but her tone of voice, indicates she is angry, and her eyes indicate that she is sad, requires the kind of selective choice abilities of the OFC. Even a young child can learn not to run into the street, because the AMG is fully developed by that age. But to make sense of which facial signal is the relevant one to respond to can be quite confusing for a young child, because the OFC is immature. In fact the OFC along with the whole PFC does not fully mature until well into the 20's. The late maturation of the PFC explains why children are poor at reasoning, judgement, planning and affect regulation.

6. 2. THE RESPONSE OUTPUT PART OF EMOTION

Once an appraisal occurs, the AMG and OFC signal the HYPO and BS to activate an emotional response, consisting of three objectively measurable 'biologic' parts, and a 'subjective part'. The biologic responses involve the body's *biochemistry, autonomic nervous system* and *motor behavior*, which can all occur without conscious awareness. The subjective part is conscious and is called *feelings*. It is presumed that the biologic changes contribute to the subjective feelings, but the exact mechanisms have not been fully worked out yet.

6. 2. 1. Biochemistry

Neuroepinephrine [NE] is the first initial biochemical response activated by the emotional system (LeDoux 1994, 1995). The NE response is an 'arousal' response, a rather non-specific reaction to any kind of emotionally significant stimulus. NE causes an 'alerting and orienting' to the source of the stimulation, enhances attentional focus to the relevant stimulus and makes the brain more ready to receive incoming stimuli. For example a herd of animals all perk up their ears and turn them in the direction of a sudden sound. A crowded auditorium will all turn to the direction of a sudden noise. NE also increases activity of the sympathetic nervous system (SNS) to dilate the pupils, increase heart rate and blood flow to the muscles. Finally NE activates the AMG to signal the hippocampus [HIPPO] 'This is significant. Remember it!'

Cortisol, the stress hormone, is released after NE, if the situation is stressful. The HYPO activates the production of cortisol, via a long and complex sequence of events, referred to as the hypothalamic-pituitary-adrenal axis [HPA axis] (Panksepp 1998). The HYPO produces cortisol-releasing-factor [CRF] which signals the pituitary to produce adrenocortico-tropic hormone [ACTH], which signals the adrenal gland to produce cortisol, which then disperses through the blood stream and can affect all parts of the body. Production of cortisol is very well regulated. In a delicate *negative feedback loop*, cortisol is detected by cortisol receptors in the brain, most importantly in the HYPO, HIPPO and AMG. This negative feedback system signals the HPA axis to shut off the production of cortisol.

Cortisol helps an animal *cope* with stress. Cortisol provides the kinds of physiologic changes in the body necessary to support coping behaviors, such as running or, fighting. Specifically cortisol increases blood pressure and increases blood glucose to make more glucose available to muscles for energy. Cortisol is also part of a feedback system for NE. Cortisol acts to shut off the release of NE.

The HPA axis and the cortisol respone must be in the optimal range. Hyper-reactivity of the HPA axis, can occur as a result of *early childhood truama, prolonged separations and neglect* (Teicher 2000; Yehuda 2001). Very high levels of cortisol are harmful. Very high levels of cortisol, can atrophy or even destroy cells in the HIPPO (Bremner, Randal et al. 1995; Sapolsky 1996). Such damage, can lead to impairments of memory. Since cortisol receptors in the HIPPO help shut off cortisol, damage to the HIPPO can contribute to excessively high cortisol levels. The damaging effect of cortisol on the HIPPO helps explain why with trauma there may be reduced rather than increased memory for the traumatic situation as one would expect. Indeed what we often find in psychoanalytic patients who have been traumatized, by physical or sexual violence, their memory is spotty for childhood events. This may seem like repression, or dissociation, but may in fact be the result of damage to the memory system itself. Also because damage to the HIPPO can impair cortisol regulation, these patients often react, even to benign situations, as if they are overwhelming and stressful. Chronic hyper-reactivity of the HPA axis, predisposes these individuals to depression and anxiety, presumably due to chronic high levels of CRF (Silk 2000; Yehuda 2001).

On the opposite end, *too little* cortisol can be harmful. For reasons not yet understood, the neurobiologic profile of individuals who suffer from post traumatic stress disorder, PTSD, appears to be – hyper-reactivity of the HPA axis, with high levels of CRF but a *blunted* cortisol response, – i.e. the cortisol rise during traumatic or stressful events is less than expected . This profile, may be the factor that predisposes individuals to develop PTSD and can explain the often contradictory symptoms of PTSD. With a blunted cortisol response, NE arousal does not shut off. PTSD patients show hyper-arousal, vigilance, increased startle (e.g,, alerting and orienting response), and heightened memory for the traumatic event in the form of flashbacks[6] (Van der Kolk 1993). The part of PTSD characterized by emotional detachment, may be due to the low cortisol itself. Low levels of cortisol have been found to be associated with impaired transfer of information from the right hemisphere (associated more with emotional processing) to the left hemisphere (associated more with verbal processing). Consequently these patients appear emotionally detached, because they can't verbalize their feelings (Henry 1993). An unfortunate byproduct of this reduced expressiveness is that these individuals may be unable to adequately use other individuals to help them reduce their distress. One woman patient, a graduate student, suffered neglect, because her mother was chronically depressed. She suffered trauma, from her parent's talk about their holocaust experiences and because she was sexually molested at 8 years old, by a male cousin. She suffered flashbacks, felt chronically stressed, and overwhelmed, but she was completely detached from any awareness of emotional distress about what had happened to her as a child. She kept her distress to herself most of the time. In the transference, she was very warm and accepting of me. If I did anything wrong, or made any mistakes she attributed it to how many responsibilities I had manage and was forgiving because I helped so many people. She had no ability however to forgive herself, or give herself any warmth or empathy for what she had been through and had to cope with as a child, because she could not get in touch with the idea that anything terrible had happened to her. This kind of emotional detachment is problematic in the psychoanalytic treatment situation. She could not feel any awareness of transference feelings. As is well known, unless a person can *feel the emotion* within the transference, (i.e., fear, or dependency), interpretations will be ineffective.

Endorphins: Endorphins, the brain's endogenous opiates, are hormones, released by the HYPO, as well as other brain regions, which reduce pain (both physical and emotional), create a sense of well being,

[6] Memory for events normally goes through, at least, some degree of fading. In the case of PTSD, instead what occurs is memory does not fade, and the person has flashbacks- the intense, vivid recall of traumatic events that are triggered very easily and do not appear to fade.

and promote mother-infant attachment (Kalin, Shelton et al. 1995; Panksepp 1998). Because endorphins are released during pain, individuals can become 'addicted' to practices such as cutting themselves.

Mother-infant attachment involves two emotional processes, *separation distress* and *reunion comfort*. Each of these emotions involves different brain chemistry (Hofer 1996; Meaney 2001; Panksepp 1998; Suomi 1999). *Separation distress* is mediated by the HPA axis. During a separation, CRF is released, leading to increased cortisol. When the baby is reunited with the mother, *reunion comfort* is mediated by endorphins. Reunion with the mother increases endorphin levels in both the baby and the mother! By shutting off cortisol, endorphins shut off the separation distress response. The endorphins give both baby and mother a feeling of well being and decreased emotional pain. Babies, in a sense, become addicted to their mothers. This may help explain the insatiable need of certain patients to find a mother figure in the analyst. But mothers also gets 'addicted' to their babies as well, which may underlie what Winnicott calls Primary Maternal Preoccupation {ref}.

Experiements on monkeys (Kalin, Shelton et al. 1995) where endorphin levels can be easily manipulated, illustrate a number of clinically relevant points. When endorphins are pharmacologically blocked, in the mother and infant, both engage in more intense reunion behavior, (which consists of a kind of hugging and vocalizing), ostensibly to increase their endorphins. If on the other hand, exogenous endorphins are given in the form of morphine, both decrease their reunion behaviors, to down regulate endorphin. This data suggests why individuals with early separations and loss may fail to have adequate endogenous endorphins and may be more susceptible to opiate addiction, and more susceptible to excessive dependency in the transference situation. This data also explains why heroin or morphine addicts make such poor parents. They do not react properly to the infant's need for contact.

Oxytocin is a hormone, released by the pituitary gland, during reproductive behaviors such as, sexual activity, labor and nursing (Insel 1997). The oxytocin released during sexual acitivity enhances monogamous pair bonding between mating partners and parental care toward offspring. If the action of Oxytocin is blocked, both monogamy and parental care are impaired. What this suggests is that sexuality in humans serves not just reproductive functions, but enhances the feelings of attachment between the partners and also enhances their desire to care for their children. We observe this clinically when couples are not having sex, they are very often emotionally distant from each other. As analysts we tend to think of the problem as one in which their emotional conflicts are getting in the way of a good sexual relationship. And while this is definitely an important factor to consider, it also appears that the sexual connection itself, serves an attachment function, and may be what helps partners feel close and committed to one another.

6. 2. 2. Autonomic nervous system

The Autonomic Nervous System [ANS] carrys sensory and motor information between the brain and the internal viscera (Schore 1994). The HYPO is the brain center which regulates the ANS. The ANS is divided into the Sympathetic [SNS] and Parasympathetic [PNS] nervous system. ALL organs of the body are innervated by branch from the SNS and PNS. The SNS and PNS each have an opposite effect. The SNS regulates body organs to cope with external stress, and the PNS regulates body organs to relax, regenerate and repair. The SNS is often referred to as the *Fight or Flight* system and the PNS as the *Rest and Digest* system.

Much of the emotional response is the result of ANS activity. For example with fear or anxiety the heart races, the blood vessels pump, the mouth turns dry, the stomach slows down (which may cause nausea or decreased appetite), and the large intestine may speed up (which may cause cramps and diarrhea). Much of what we show in our facial expression of emotion is the result of ANS activity. The face flushes with embarrassment, or shows pallor with emotional 'shock'. The pupils dilate with fear. The eyes tear with sadness. Even our voice shows emotion as a result of ANS activity. The vocal cords quiver with fear and anxiety. Disgust, as an emotion, most likely began as a way of getting rid of spoiled food. When disgust is directed, not at a food, but at a person's behavior, our ANS may activate in the same way, with a slowing of the stomach and even some reverse peristalsis.

Somatizing and Alexithymia Neuroscientists reserve the word 'feelings' for the conscious subjective experience of emotion. Some individuals only experience the 'somatic' aspect of emotion. These individuals fail to feel emotions *as subjective feelings*, and have difficulty putting their emotions into words. A person who 'somatizes' may complain of headache, or dizziness, or stomach pain, or 'a lump in the throat' or chest pain, but will not connect these somatic complaints with feelings of fear or anxiety. It appears clinically that such patients are often poor analytic or even psychotherapy candidates because they have difficulty reflecting on their subjective emotional state and talking about feelings. This condition has been called Alexithymia (Krystal 1974). Education is helpful to enable these patients to engage in psychotherapy. Education consists in explaining that all emotion has a somatic part; and that what they are experiencing in their body is the result of emotion (Taylor 1992). They need to *re*-interpret their physical symptoms as an emotional response and to put words to those emotions. It appears that the process of *putting words to emotions* helps shift the experience away from the body and into subjective feeling states where they can be talked about. Clinically we all observe that simply talking about upsetting feelings can be calming and helpful, even if no 'analytic interpretation' occurs. It may be that talking, engages conscious self reflection, represents experience at the symbolic abstract level, and activates the highest levels of the brain, the PFC. The orbito-frontal region of the PFC (the OFC) is the highest center of affect regulation and is crucial in being able to experience situations from alternate perspectives.

Anxiety Disorders: The emotions of anxiety and fear have a similar biologic profile (Davis 1992). When the AMG appraises a stimulus as dangerous or associated with danger, (in the way that the sound of a snapping twig can be associated with the approach of a tiger), the AMG can trigger fear or anxiety. The AMG signals the HPYO and BS to produce a whole host of body responses. NE is released producing vigilance. SNS activity produces sweaty palms, difficulty swallowing, heart palpitations, light headedness or dizziness, difficulty breathing, queasiness and erectile difficulty in males. Increased tension in skeletal muscles to enhance running or fighting, can lead to complaints of chest pain, from tightening of the intercostals muscles, and headache, from tension in scalp muscles. In fact what many 'somatizers' report as physical symptoms, are essentially, the physical accompaniments of anxiety and fear.

Panic attacks are a severe form of anxiety. In panic, the patient has a sense of dread or impending doom and difficulty breathing. Neurimaging studies using fMRI show increased activity of the AMG.[7] Since even

[7] FMRI is a neuroimaging device which measures blood flow to various brain regions. Blood flow brings in oxygenated hemoglobin and serves as indirect measure of brain metabolism. Increase blood flow to a region, indicates increased metabolic acitivity of that region. Neuroscientists use this kind of data to localize what brain region is involved with a particular mental process. For example on memory tasks the fMRI shows increased activity in the HIPPO.

the physiologic changes associated with anxiety, such as increased heart rate, can be unconsciously appraised by the AMG as 'danger signs', even low levels of anxiety can escalate into panic states.

The treatment of anxiety follows from the biology. Pharmacologic treatments involve benzodiazapines [BNZ] and selective serotonin reuptake inhibitors [SSRI's] both act either directly or indirectly to inhibit the reactions of the AMG. Cognitive Behavior Therapy [CBT] focuses on the appraisal aspect of emotion – using voluntary conscious appraisals to counteract automatic non-consciously derived ones. This engages the PFC which has more ability to select a more adaptive response and inhibit an automatic maladaptive one (Panksepp 1998). For example, an anxious person can become more anxious because their appraisal system signals 'Danger! Heart is racing'. Patients can be instructed to notice heart palpitations and consciously assign them a more benign appraisal value. "When my heart rate goes up it just means I am anxious. It does not mean I am having a heart attack!" Although many analysts do not recognize it, *CBT-like* techniques are used in transference interpretations. I would often say to my patient the professor, who felt his mother was cold and did not love or care about him. She could emotionally 'ice' him out. Once she locked him out of his house by his mother, when he was 10 years old, because of a prank in which he scared her with a *plastic* spider. "When I don't talk, it is as if you assume I don't care about you; I have locked you out. But it could mean I am intently listening and do not want to interrupt what you are saying" A more benign appraisal can keep anxiety at more manageable levels. Since severe anxiety interferes with memory and cognitive functioning, reducing excessive levels of anxiety, in the case of analysis, promotes greater tolerance for the treatment and more ability to reflect on the process.

Chronic Stress. Stress is related to situations of danger and threat, and can either be acute (for a short time) or chronic (for a long time). An acute stress response is healthy and adaptive because it enables the individual to cope with the stress producing situation. Chronic stress however can be harmful both physically and emotionally (Sapolsky 1996). With chronic stress, the initial acute phase does not return to baseline 'homeostatic' levels. During during chronic stress, there are chronically high levels of SNS activity, CRF and cortisol, which can potentially cause physical and emotional illnesses, or exacerbate existing ones – including, hypertension, heart disease, an altered inflammatory or immune response, asthma, bone resorption and osteoporosis, Diabetes Mellitus, Depression Anxiety (Nemeroff 2002). Chronic stress in *early life*, i.e. childhood neglect and abuse, (physical and sexual) can lead to *life long, permanent* hyperactivity of the HPA axis (Meaney 2001; Newport, Stowe et al. 2002). When these children grow up they are more likely to develop depression, and anxiety as well as many of the physical conditions mentioned above.

6. 2. 3. Motor behavior

Motor behaviors which result from emotion are generally activated involuntarily and without conscious awareness. The most common emotional behavior is the 'freeze response', the sudden stillness that occurs as an animal first detects danger (i.e., the 'deer in the headlights' reaction). Others include running, hitting, hugging, waving, facial expression and changes in posture. A tall erect stance is associated with pride, while a 'slumped down' posture is associated with sadness. The facial expressions of emotion involve movement of the muscles of the face. Different muscles groups of the face are contracted to produce frowning (anger), smiling (happiness) raising the eye brows (surprise), curling the lip (disgust). Contraction of the eye muscles alters direction of gaze. Eyelids and eyes are lowered in shame, staring straight ahead is associated with anger, gazing into the eyes of a loved one is associated with love.

Each of Ekman's (1983, 1990, 1993) six 'basic emotions'– anger, fear, sadness, disgust, happiness and surprise – has a characteristic facial expression, specific muscle groups that contract in the face and a characteristic profile of ANS activity that corresponds with each emotion. The biochemistry, ANS activity, motor behavior (i.e., facial expression) and subjective feelings of emotion are linked. In experiments, subjects either see emotion laden images, or imagine an emotion or are taught to contract the muscles groups of their face which correspond to an emotion. ANS measures are taken, such as heart rate or skin temperature. When subjects either watch images or imagine a emotion, their ANS shows the characteristic profile. But also, many subjects who *simply contract the appropriate facial muscles*, even though they do not know what emotion this corresponds to, also have the ANS and *subjectively feel* the feeling!

When two people converse, they automatically and unconsciously match each other's non-verbal behavior, including facial expression, gesture and body posture. Studies show that the degree to which people match correlates with their sense of rapport and liking each other- even though the process of matching is entirely outside of conscious awareness. In an experiment by Feldstein & Welkowitz (1978) two complete strangers are asked to talk with one another and afterwards each is asked to rate whether or not they would be interested in going out for dinner with the other person. Their conversation is recorded on videotape. When there is a high degree of matching within the dyad, they are both more likely to want to meet up with each other again, than in dyads where matching is low. Studies between hospitalized patients and therapists indicate that when the patient and therapist are high in matching, the therapist is more likely to have a positive feeling toward the patient. It is theorized (Beebe & Lachmann 1988) that matching contributes to empathy. Since a person's facial expression is linked to a characteristic pattern of ANS activity, when we match the facial expression of another, we set up within ourselves the same body state that they have…and we literally can *feel what they feel.*

Within a dyad there can be *complimentarity*, in which each member displays a *complimentary* behavior. In matching, behavior is *similar*. When the baby coos, the mother coos. In complimentarity, if the baby cries, the mother picks up the baby and soothes it, i.e., she does not cry herself. There are a number of implications of matching and complimentarity for psychoanalysis (Beebe & Lachmann 1988). Such processes occur without conscious voluntary control over the behavior. One matches or reciprocates automatically and without any thinking or planning involved. Matching may explain why the analyst often feels they know what is going on in the patient, simply from seeing them in the waiting room. Once the patient is lying on the couch, the main source for non-verbal matching, i.e., the face, is lost. Except for vocal quailities such as tone, timbre, rhythm and pitch, other sources are diminished, such as body movement and posture, although more so for the patient than analyst. The decreased ability for non-verbal matching, albeit that it occurs outside conscious awareness, may explain why some patients on the couch have such a hard time feeling the analyst is 'there' in a caring way.

The treatment of a professional woman in analysis, illustrates matching and complimentarity (Pally 1996, 2001). During her childhood, she suffered many long absences from her depressed mother, who was hospitalized on several occasions. The woman had two different ways of talking. She spoke like a competent, independent minded and self sufficient professional eager to work analysis. At other times she seemed small and fragile, anxious about beginning analysis, for fear of the unknown, of getting hurt, or of discovering something frightening.

In the first several months of the analysis, I began to notice myself talking in a rhythmic, soft, high pitched voice as if I was a mother with a baby. In the transference she began to feel completely protected by me, as if in a "safe bubble". In the countertransference, I felt like a mother in a state of "primary maternal preoccupation" (Winnicott 1956). I was matching her *the little baby vocal rhythms and pitch,* with my *mothering vocal rhythms and pitch*, and I was providing the complimentary response to her distress with rocking motion to comfort and soothe her.

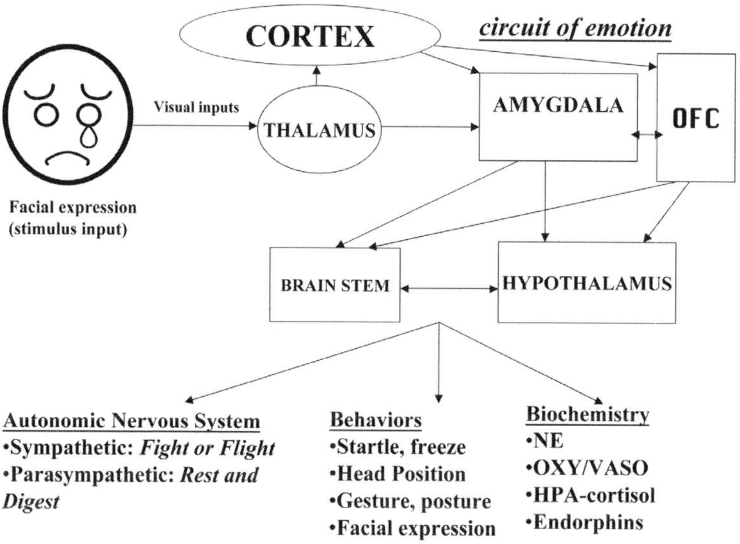

7. CLINICAL IMPLICATIONS

7. 1. EMOTION KEEPS US ORGANIZED

We react emotionally to the environment, because emotion organizes the brain and body to perform the behaviors necessary to cope with environmental situations. We must react emotionally, for physical survival, and, because we are such highly social creatures, for *surviving within our social network*. For example, shame is generally induced in a person, when their exploratory behavior or exuberance is abruptly interrupted (Schore 1994). Shame can occur when a child is chastised by its mother for climbing up on a shelf to see what is there, with "STOP! You will FALL!" or for running into a room full of relatives to tell mother about what happened at school and being suddenly told to "Stop! Please greet Auntie Jane!" The emotion of shame 'organizes' the necessary adaptive response to these abrupt prohibitions. Shame is designed to inhibit exploratory behavior and dampen enthusiasm. Shame involves increased activity of the PNS (ie, which slows the heart rate, decreases blood flow to the muscles) to support the process of withdrawal and behavioral inhibition. Behaviorally, the child stops their behavior and dampens outward signs of enthusiasm, i.e., lowers the head and eyes, and may even withdraw or hide. Interestingly, shame does not develop, until the second year of life, perhaps because this when when the child develops sufficient locomotor abilities and the motivation to exuberantly explore its environment. One way to look at the adaptive value of shame, therefore, is that shame helps us to learn not only rules of safety, but also, the *rules of society*. Each social group and culture has its own norms. Adaptive functioning requires us to learn these rules. Those that don't are marginalized. We learn these rules which govern our social behavior, as a type of procedural memory; meaning that once it is learned it can be activated unconsciously and automatically (Grigsby 1994).

7. 2. Emotion must be regulated

Emotional regulation means two things. One, we REACT emotionally and RECOVER afterward, i.e., return to baseline. Two, emotion occurs within an 'optimal range of intensity'. Individuals who are not well regulated often suffer physical and psychological conditions, such as somatic symptoms, anxiety, depression and 'character' traits.

Shame, guilt, fear and anxiety are all good examples of how too little emotion or too much emotion is not good. Sufficient shame is necessary for learning the social norms of how to relate to others. Therefore, civilization' depends, on shame as much as it does on guilt. In the clinical situation we often see patients who don't 'read' or respond properly to social boundaries. Some are too intrusive, as occurs with borderline patients. Some are too demanding of attention, as occurs with narcissistic conditions. These individuals have difficulty with friends, family and colleagues. In the transference, the intrusive ones make us feel invaded, the attention seeking ones, make us feel used. Perhaps these individuals were raised by parents who did not provide appropriate 'shame guidance', to learn the proper social rules regarding distance and attention.

Perhaps these individuals grew up with parents who did not know these rules and thus passed these traits on to their children, as a form of procedural memory. On the other hand, *excessive* shame responses are not healthy. Some types of narcissitic patients have excessive shame and they are inhibited in social situations. The model proposed by Kohut (Kohut 1984) it may be that narcissistic pathology develops when shaming experiences are too harsh. The *sense of self* feels diminished and the child's enthusiastic exploration is inhibited. These individuals defend against their shame, by disavowing or dissociating from it. As a result they are both over inhibited in some arenas social interaction and under inhibited in others. Although the person's longing for expansive expression of interest and enthusiasm remains *split off*, in the treatment it emerges in the transference.

Too much guilt is obviously not adaptive, and is essentially the basis for much of psychoanalytic treatment, most notably in the treatment of the classical neurosis. Guilt, involves an increase of SNS activity, as measured by the Galvanic Skin Response (an indicator of increased sweating and the basis of the 'lie detector' test). In some respects guilt is associated with fear and anxiety and is adaptive because it aids the individual in anticipating the negative consequences of their actions. Sociopaths, and certain brain damaged individuals, appear to have too little guilt. They lack the normal GSR response to situations of danger, and thus do not experience guilt (Damasio 1994). Sociopathic individuals are very poor psychoanalytic patients. They do not experience the necessary anxiety about the negative consequences of their behavior, even at an unconscious level. It appears that due to the lack of GSR response, maladaptive, even dangerous behavior remains ego syntonic and unlike the borderline patients whose ego-syntonic behavior is treatable, sociopaths lack the necessary bio-phyisology to shift over to an ego-dystonic perspective.

All analysts are well acquainted with patients who appear to be in a state of vigilance, on 'high alert' so to speak for danger, always perceiving other people as threatening, even the analyst who may never have said or done anything harmful. The neuroscience perspective helps explain such a transference. When a person is anxious or fearful, from whatever cause, a recent rape or a childhood trauma, or even an inborn temperamental trait, the brain's appraisal system is more likely to appraise 'in coming stimuli' as dangerous, leading potentially to the vicious cycle in which fear breeds even more fear. This is what typically occurs with PTSD. For example a physical assault or accident, generates fear. Once the person is afraid, their appraisal system is biased toward danger and they can develop a state of vigilance, in which they are more likely to interpret situations as dangerous. The most commonly known example, is the war veteran who shakes with fear to loud noises such as the back firing of a car. This is one of the best uses of neuroscience I have found. Trauma patients benefit from understanding this kind of cognitive bias, and can use it to re-interpret what they consider dangerous. Even in the transference, patients who have suffered trauma in childhood are on high alert to the dangers of being abandoned, or abused, and may perhaps startle, cringe or cower to otherwise neutral movements by the analyst – a shift in the chair, a tap of the foot, even merely sitting behind the patient.

But too little fear and anxiety is not good either. In the case mentioned earlier of the woman professional, she disavowed fear. Because of job responsibilities she was often required to travel to dangerous areas, where she would walk alone at night on isolated streets. Also as a result of her job, she was once stalked and threatend by a former employee but did not think of letting the authorities know. In the transference she said she was in a "protected bubble" with me. I would interpret to her that while such safety must be a good feeling, that without access to her own fear, she was not able to properly protect herself from danger. She replied that to deny fear, was a way of maintaining a tie to her mother even in the mother's absense. For her it was as if, mother would be there and save her. If she accepted fear, it would mean, she would have to experience being alone, without a mother who would come and rescue her.

7. 3. Why we show our feelings

One reason we *show* our feelings, is simply to communicate to others what we are feeling. A bright smile can communicate to others that we are happy, tears that we are sad, in much the same as words communicate. But another reason, beyond communication *per se*, is that we are innately designed such that the expression of emotion by one person automatically and unconsciously *regulates* the behavior, emotion and body physiology of the person to whom it is expressed! This kind of mutual regulation occurs, because the non-verbal behaviors of emotion of one person, are assessed by the other's AMG and OFC appraisal centers, and in turn (through connections with the HYPO and BS) generate biochemical changes, ANS activity, and behaviors in the other. There are many examples of, how mutual regulation through the non-verbal behavior of emotion, is evident, through out life. The baby's cry of distress, is not simply to let mother know how the baby feels, it is biologically designed to activate the very nurturant care taking responses, such as hugging, holding and touching which will relieve the distress. This has been well studied in many animal models (Hofer 1996b; Kalin 1993; Kalin, Shelton et al. 1995) and is presumed to operate in a similar way in humans. When a child is too dampened down with shame for an extended period, this is designed to activate the parent to re-animate the child, with words, and perhaps touch, of encouragement or forgiveness (Schore 1994). Even between adults, when a person hangs their head in shame, this can serve to activate forgiveness in the other. As two strangers approach, if one stares directly into the eyes of the other, this is a sign of threat and the other backs off. If however, as they approach, one averts their eyes, this signals they are not a threat and that the other can approach. When a sad person cries, this activates sympathy in others. The clinical implication is that the patient's non-verbal behavior will activate these 'emotion-regulation' responses automatically and involuntarily in the analyst as well. The analysts own self reflection, as to how they feel *inclined* to respond, may be an indication of the emotion in the patient, even if the patient is not conscious of that emotion. The analyst's sensitivity to the patient's silence – *how long to wait, when and how to intervene* – probably derives from these kinds of non-verbal cues.

In *certain respects*, the neuroscience perspective supports the idea of projective identification. However, unlike the psychoanalytic concept, the biologic system of matching and empathy is innate, and automatic. It does not imply any *defensive* purposes, although, most likely this system can be used defensively. Whether the patient's non-verbal emotion behavior is a way to defend against conscious affects, or simply a reflection of those affects, it can activate "countertransferense" feeling states in the analyst that can be utilized to formulate an interpretation. One male patient, an attorney kept pressing me to have a sexual relationship, often trying to humiliate me for not being willing to have such an experience (Pally 2001). I had a pronounced feeling of disgust rather than feeling either flattered or sexually aroused myself. He would spout political and moral ideals and aggressively try to get me to agree with his ideas, in this case demeaning the psychoanalytic technique which prevents me from agreeing. I found his ideas more repugnant than agreeable. Aware that there was a lot of hostility in the transference, toward me as a woman and as an authority figure, and informed by my feeling of disgust, I finally interpreted, "It is as if you want to shove your ideas down my throat and make me swallow them even if I don't want to." Another male patient, a magazine editor, feels he deserves nothing and will aggravate people if he asks for help or support. He is always cheerful and engaging. He talks about his concerns as if he is telling an interesting story. Despite his charming flare for entertaining me, I continually feel inclined to reassure him, to offer words of encouragement and hope. It is as if, he activates 'comforting behavioral' responses in me, *somehow* by his non-verbal manner. While it could be argued that my need to comfort could reflect my own 'issues' so to speak, I think these kinds of inclinations in therapists are also part of the natural and normal way we as humans are designed to react to one another's distress, even if only expressed non-verbally and maybe even outside conscious awareness, and that analyst's can use these feelings in themselves as guides for transference interpretations.

One theory about why individuals with alexithymia often suffer dysregulated emotion is that they are not only poor at verbally expressing feelings, they are also poor at expressing the non-verbal behaviors of emotion. Therefore they are less likely to activate the kinds of responses in others which will serve to regulate them. For example even if they are quite sad or frightened, they may talk in a monotone with a bland face. For the therapist this can lead to boredom. Without the outwardly expressed non-verbal behavior of emotion, the therapist is not activated to either empathize or sympathize. The therapist neither 'gets' what the patient is feeling nor tries to comfort and soothe.

7. 4. Why we consciously feel our feelings

The vast majority of brain processes are *non-conscious*, which includes the psychoanalytic concepts of unconscious (both the conflictual and conflict free aspects), as well as the pre-conscious. Only a very small percentage ever becomes conscious. Consciousness occurs 'after the fact'. First something is processed non-consciously and then, a few milliseconds later, can become available to consciousness. With respect to emotion, the brain first generates emotion non-consciously, and only afterward do we become aware of the emotional behavior and conscious awareness of a 'feeling'.

But what is the purpose of consciousness? This is as yet not well understood. One theory is that consciousness plays a role in the brain's oversight system (Posner and Rothbart 1998; Freeman 2000; Fuster 2003). In other words, consciousness may play a *monitoring role*, for determining whether a particular emotional response was adaptive. If an emotional response becomes conscious and the person recognizes the emotion as inappropriate the brain can 're-appraise' the situation and switch to a different emotion. Or the brain can inhibit the 'inappropriate' behavior which follows from that emotion and voluntarily initiate a more adaptive one. For example, I often find myself feeling tense when a patient has a long silence. I notice myself *start* to speak, *subvocally*. After I am aware of my mouth starting to move, or perhaps my body shift slightly, but before any words are actually uttered, I will realize I am responding more to my own tension, than to what the patient might need from me, and I inhibit my talking. In the case of one male patient, the magazine editor, whose transference involves a very high degree of wanting to meet my needs, he is hyper-alert to sensing these tiny movements of mine, and often, particularly in the beginning of the analysis, would stop his internal reverie and says to me, "Go ahead, say what you were going to say. It is alright." Now, well into the analysis, many times he is able to ignore my movements and to continue the silence or begin talking *according to his own need.*

Much of the time, emotion is activated entirely outside of awareness. This means we are not aware of an emotional 'feeling' and may not even be aware of any physiologic changes in our body, or even our motor actions, such as facial expression. Clinically this is significant, because without recognizing it, both patient and analyst can still effect each other's affect, physiology and behavior, albeit non-consciously.

The value of being conscious of feelings, particularly the uniquely human type of consciousness known as 'self-reflective consciousness', is that it appears to facilitate regulation of emotion. This is particularly relevant to psychoanalysis, since putting feelings into words, (a major part of the analytic treatment), enhances self-reflective consciousness. Humans, like all mammals, have the innate ability to be consciously aware in the moment. Only humans can be *conscious of being aware* and, can reflect on their experience (Edelman 1992; Edelman & Tononi 2000; Damasio 1994). Such reflection helps an individual to recognize what aspect of the situation is stimulating their emotional response, and that there are a number of alternative ways of responding. Babies are born with basic 'animal' conscious awareness. As they develop linguistic ability, they can reflect more and more on their conscious awareness. Such reflection makes possible re-appraisal of a situation and the recognition of alternatives to one's initial response. I would argue that by enhancing self-refletive consciousness, verbalizing one's emotion, serves to enhance self-monitoring and to recognize whether one's initial response was appropriate and adaptive and to inhibit their behavioral response and try to take in more stimuli to see if it is possible to shift to another emotion with a different behavioral response. In analysis, by verbally expressing emotions, the patient is more able to reflect, rather than act out; more able to move outside of the regressed transference response and take a more mature stance; more able to utilize the observing ego, rather than just the experiencing ego. This helps explain why, even without any interpretive intervention, eventually most people feel better for having talked about their emotions. In some cases reflection on feelings helps patients to actually *feel differently* about a situation.

However, in many cases, the feelings are too engrained, too deeply rooted, and no amount of verbalizing and self-reflective consciousness can alter this. I think one of the most valuable lessons of neuroscience is that we cannot control what emotions we have. Emotion is activated involuntarily and out of our awareness, and that there is nothing maladaptive about any particular emotion per se. What *can be maladaptive* are the behaviors which result from an emotion. Conscious reflection on emotion, which is enhanced by putting words to emotions, even if it can't change the emotion, can help inhibit the maladaptive behavior and activate a more adaptive one, because, verbal expression and conscious reflection engages the PFC, the brain's center for abstract representation and selevtive choice of behavior (Fuster 2003) I make this kind of perspective explicit with patients. I tell them that, particularly with very painful experiences, such as occurs with the extremes of emotional and physical abuse, the treatment may not help them feel any less hurt, any less ashamed, any less angry (although I hope it does). It can however help them not act on those feelings. In the long run, by not acting, and by tolerating the feelings more, in fact they are morel likely to dissipate on their own (as emotions naturally do), and thus, in the end, feel less intense and painful.

7. 5. Memory and Defenses

The encoding of the long term memory for events, generally referred to as 'declarative' or 'explicit' memory, is strengthened by *optimal* levels of emotion (McGaugh 1995). When an event is appraised as 'emotionally significant', an increase of the neurotransmitter NE is released in the brain, which activates the AMG to send a signal to the HIPPO to encode the memory. Conversely, when not enough emotion is generated, not enough NE will be released and long term memory for the event is only weakly encoded or not at all encoded. Once memory is encoded, it must be retrieved for us to be conscious of it. The stronger the encoding process, the more likely the memory will be able to be retrieved.

Too much and too little emotion actually impairs memory. At one end of the continuum, too much emotional arousal can interfere with the encoding of explicit memory. In this case the hormone 'cortisol' is involved. When cortisol gets very high, this can cause impairment or even cell death in the HIPPO. For example veterans exposed to war violence and children exposed to physical and sexual abuse show a smaller than normal size HIPPO (Bremner, Randal et al. 1995; Sapolsky 1996). What this implies clinically, is that many patients who appear to have 'repressed' their traumatic experiences, may in fact suffer from a biologic impairment of memory encoding.

At the other end, too little emotion can serve the purpose of impairing memory and recall of painful aspects of a situation (Edelman 1992; Fuster 1996; Edelman and Tononi 2000; Fuster 2003). Some patients utilize this for defensive purposes. One woman in treatment, a business analyst, claims that she had a happy childhood, but can't remember much of it. In the transference she seems to often feel criticized by me, but if I make a comment about it, she says "No I wasn't feeling that way". This woman uses the defense of disavowal of affect. Characteristically she might say "As a child I never could please my mother, but I didn't let it effect me". One way to understand her lack of childhood memories is that when a person ignores or minimizes the emotional impact of an event, this can impair their ability to *store* the event in long term memory in the first place, or to later recall it from long term memory.[8]

As important as it is to remember, it is equally important to forget! In fact, there are brain systems for forgetting. Under the usual circumstances, all memories gradually fade to *some degree*. Insignificant ones may fade away completely. For example, most likely, you will not remember tomorrow what you had for breakfast yesterday. Where as, significant ones may simply fade in intensity. This is why you no longer

[8] From a neuroscience perspective, what analysts refer to as repression, is the inability to retrieve memories that have been encoded.

can remember the *intensely stabbing* pain you felt when you broke your leg several years ago.

Impaired 'forgetting' is one way of understanding the condition known as Post Traumatic Stress Disorder (PTSD) (Van der Kolk 1993). In PTSD the traumatic memory does not gradually fade. It remains vivid, as if occurring 'right now'. The biochemical response associated with PTSD helps to explain this disorder of 'forgetting'. CRF is released by the traumatic event, but, for reasons not yet fully understood, the cortisol response is blunted (i.e., not as high as would be expected for that level of stress). Since cortisol acts to shut off the initial NE response of emotional arousal, patients with PTSD are in a chronic state of high NE and SNS arousal, with anxiety, insomnia and hyperactive startle response, as well has highly vivid traumatic memories, that do not fade. In fact each time they 're-remember' the trauma in flashbacks, it becomes even more strongly encoded. The blunted cortisol found with trauma, may also be associated with impaired transmission of information from the right hemisphere, which processes emotion, to the left hemisphere, which processes language. These patients have emotion, but may appear emotionally detached, because they cannot verbalize what they feel. To treat the PTSD, it is necessary to help them experience their feelings and put them into words, which enables the traumatic memory to be integrated with the whole body of prior memory. In this way the memory *fades into the past*.

Defenses are mobilized to avoid painful affects. Research suggests a similar mechanism between managing physical pain and managing emotional pain (Sierra and Berrios 1998; Simeon, Guralnik et al. 2000). Pharmacologic studies and neuroimaging studies illustrate how the limbic system, may be employed to reduced physical pain and to activate psychological defenses. In the pharmacolgic treatment of pain, anesthetics and analgesics are used. Anesthetics, (e.g. Novocain), block all pain sensation. With analgesics, such as the narcotic morphine, the pain is still there, but it doesn't bother the person. It is known that analgesics reduce emotional pain as well as physical pain. A region of the limbic system, known as the Cingulate Cortex, is involved in pain awareness. Subjects are hypnotized and given either a suggestion for pain anesthesia or pain analgesia.. . If a person is given the hypnotic suggestion for anesthesia, neuroimaging reveals decreased activity in the cingulate and decreased in the *somato-sensory cortex* as well. If the suggestion is for analgesia, there is diminished activity *only in the cingulated.* Activitiy is normal in the somato-sensory cortex. The cingulate diminishes awareness of the pain, even though the sensory cortex still records the pain sensations! This suggests that the cingulate may also be employed in defenses against awareness of emotional pain such as disavowal and dissociation, . The PFC may also be involved in defenses. The PFC is the executive center of the brain and functions mostly to inhibit activity in lower centers. A case of a woman who suffered an emotional shock and developed a hysterical paralysis indicates the role of the PFC in the defense (Marshall, Halligan et al. 1997). After her arm became 'paralyzed', neuroimaging revealed low activity in the right motor cortex and high activity in the PFC. Presumably the PFC was inhibiting her motor cortex. In the case of organic paralysis due to stroke, neuroimaging typically shows low activity of motor cortex, but with normal activity in the PFC. If a subject is given a hypnotic suggestion for paralysis, the same neuroimaging profile as with 'hysterical' paralysis occurs. Putting this data together, what is suggested is that the effect of emotion on the PFC, as well as hypnotic suggestion may utilize similar neural circuits and may help explain certain aspects of repression. Emotion, or Hypnotic Suggestion, acting on the PFC, inhibits activity in the lower cortical sites where the memory is stored, and therefore sites receive insufficient activity for the memory to be retrieved (see footnote 7).

7. 6. Rational Thinking is not so Rational

Neuroscientists now consider that phrases such as a "heart felt" decision or a "gut decision", may not just be metaphors (Damasio 1994). Everytime we have an experience, it activates emotion which activates the body. The OFC stores the memory of the emotion, including the body changes involved in that emotion, *in association with* the sensory, motor and contextual aspects of the experience. The body changes are referred to as the 'somatic marker' of the experience. The OFC uses the somatic markers as guides in the decision making process. When a person has to make a decision or a choice as to how to behave in a particular situation, the brain relies on how that decision affected the body in a similar *past* situation. In essence the brain relies on somatic markers to determine "How did I feel the last time I acted in such and such a way in this same type of situation" . Patients with brain damage to the OFC and many individuals with sociopathy show impaired decision making, presumably because of impairments in their ability to assess somatic markers. It is assumed that sociopaths cannot learn from the negative consequences of past behavior because of the impairment in somatic marker registration. Clinically what this suggests is that for adaptive functioning, people need to have emotions about the events of their lives, to create somatic markers of experience, because this is how we learn from past experience.

8. CONCLUSION

Emotions are not just subjective feelings. Emotions are as much about the body as is digestion and respiration. What neuroscience indicates is that these bodily aspects of emotion are very relevant to the clinical issues we address in psychoanalysis. They help to explain many of the symptoms of psychological disorders, and why they are so deeply entrenched. They also help to explain how empathy, transference, countertransference and psychological defenses occur, and even address the patient's oft asked question "why do I have to feel my feelings?"

Although emotion is now center stage in neuroscience and in psychoanalysis, there is one caveat. Some emotions are so deeply engrained we cannot change them, even with the best analysis. These patients will continue to have the old painful feelings and will have the impulse or tendency to act on those feelings. I tell patients we have no control over how we feel, only over how we act. This need not be a negative outcome however, because patients in analysis, who are able to bring their emotions into conscious awareness, can learn to inhibit the maladaptive behavior which results from that emotion.

ACKNOWLEDGMENTS

I had help in the development of these ideas from the Westside Neuroscience Study Group whose members are Hans Miller, John Schumann, Leon Sones, John Sherin, Bob Agagenian, Don Faraveau and myself; from colleagues Dan Siegel, Allan Shore, David Olds and Beatrice Beebe; and The UCLA Brain Research Institute Journal Club headed by Joaquin Fuster and Arnold Scheibel.

REFERENCES

Beebe, B. & Lachmann, F.M. (1988). The contributions of mother-infant mutual influence to the origins of self and object representations. *Psychoanalytic Psychology* 5: 305–337.

Beebe, B. & Lachmann, F.M. (1998). Co-constructing Inner and Relational Process. *Psychoanalytic Psychology* 15: 480–516.

Bremner, J.D., Randal, P., Scott, T.M., Bronen, R.A., Seibyl, J.P., Southwick, S.M., McCarthy, G., Charney, D. S. & Innis, R. B. (1995). MRI-based measurement of hippocampal volume in patients with combat-related post-traumatic stress disorder. *American Journal of Psychiatry* 152: 973–981.

Damasio, A.R. (1994). *Descartes Error*. New York: Putnam.

Damasio, A.R. (1995). Toward a neurobiology of emotion and feeling: operational concepts and hypotheses. *The Neuroscientists* 1: 19–25.

Damasio, A.R. (1999). *The Feeling of What Happens*. New York: Harcourt.

Davidson, R.J., Jackson, D.C. & Kalin, N.H. (2000). Emotion, plasticity, context and regulation: perspectives from affective neuroscience. *Psychological Bulletin* 126: 890–909.

Davidson, R.J., Putnam, K.M. & Larson, C.L. (2000). Dysfunction in the neural circuitry of emotion regulation – A possible prelude to violence. *Science* 289: 591–594.

Davis, M. (1992). The role of the amygdala in fear and anxiety. *Annual Reviews in Neuroscience.* 15: 353–375.

DeGangi, G.A., DiPietro, J.A., Greenspan, S.I. & Porges, S.W. (1991). Psychophysiological characteristics of the regulatory disordered infant. *Infant Behavior and Development* 14: 37–50.

Edelman, G. (1992). *Bright Air, Brilliant Fire*. New York: Basic Books.

Edelman, G.M. & Tononi, G. (2000). *A Universe of Consciousness: How Matter Becomes Imagination*. New York: Basic Books.

Ekman, P. (1983). Autonomic nervous system activity distinguishes among emotions. *Science* 221: 1208–1210.

Ekman, P. (1990). Voluntary facial action generates emotion specific autonomic nervous system activity. *Psychophysiology* 27: 363–383.

Ekman, P. (1993). Facial expression and emotion. *American Psychologist* 48: 384–392.

Feldstein, S. & Welkowitz, J., ed. (1978). *A Chronography of Conversation. Nonverbal Behavior and Communication.* Hillsdale: Lawrence Erlbaum Associates.

Freeman, W.J. (2000). *How Brains Make Up Their Minds*. New York: Columbia University Press.

Freud, S. (1895d) (Together with Breuer, J.). *Studies on Hysteria. S. E.* 2.

Freud, S. (1950a [1985]). Project for a scientific psychology. *S.E.* 1: 295–387.

Fuster, J.M., ed. (1996). *Frontal Lobe and the Cognitive Foundation of Behavioral Action. Neurology of Decision Making*. Berlin: Springer-Verlag.

Fuster, J.M. (2003). *Prefrontal Cortex*. Oxford, New York: Oxford University Press.

Grigsby, J. (1994). Procedural learning and the development and stability of character. *Perceptual and Motor Skills* 79: 355–370.

Gunnar, M. (1992). Reactivity of the hypothalamic-pituitary-adrenocortical system to stressors in normal infants and children. *Pediatrics* 90: 491–497.

Henry, J.P. (1993). Psychological and physiological responses to stress: The right hemisphere and the hypothalamo-pituitary-adrenal axis, an inquiry into problems of human bonding. *Physiological and Behavioral Science* 28: 369–387.

Hofer, M. (1996a). On the nature and consequences of early loss. *Psychosomatic Medicine* 58: 570–581.

Hofer, M.A. (1996b). Multiple regulators of ultrasonic vocalization in the infant rat. *Psychoneuroendocrinology* 21: 203–217.

Insel, T. (1997). A neurobiological basis of social attachment. *American Journal of Psychiatry*. 154: 726–735.

Kalin, N. (1993). The neurobiology of fear. *Scientific American* 268: 94–101.

Kalin, N.H., Shelton, S.E., Lynn, D.E. (1995). Opiate systems in mother and infant primates coordinate intimate contact during reunion. *Psychoneuroendocrinology* 20: 735–42.

Kohut, H. (1984). *How Does Analysis Cure?* Chicago: The University of Chicago Press.

Krystal, H. (1974). The genetic development of affects and affect regression. *The Annual of Psychoanalysis* 2: 98–126.

LeDoux, J. (1994). Emotion, memory and the brain. *Scientific American* 270: 32–39.

LeDoux, J. (1995). Emotion: Clues from the brain. *Annual Review of Psychology* 46: 209–235.

LeDoux, J. (1996). *The Emotional Brain*. New York: Simon and Schuster.

Marshall, J.C., Halligan, P.W., Fink, G.R., Wade, D.T. & Frackowiak, R.S.J. (1997). The functional anatomy of a hysterical paralysis. *Cognition* 64: B1–B8.

McGaugh, J.L. (1995). Emotional activation, neuromodulatory systems and memory. *Memory Distortion: How Minds, Brains and Societies Reconstruct the Past*, ed. D.L. Schacter. Cambridge: Harvard University Press: 255–273.

Meaney, M.J. (2001). Maternal care, gene expression, and the transmission of individual differences in stress reactivity across generations. *Annual Reviews in Neuroscience* 24: 1161–1192.

Nemeroff, C.B. (2002). Recent advances in the neurobiology of depression. *Psychopharmacol Bull 2002* 36 Supplement 2: 6–23.

Newport D.J., Stowe, Z.N. & Nemeroff, C.B. (2002). Parental depression: animal models of an adverse life event. *American Journal of Psychiatry* 159: 1265–83.

Pally, R. (1996). Reciprocal Responsiveness in Psychoanalysis. *Journal of Clinical Psychoanalysis. Proceedings of the New York Psychoanalytic Institute and Society* 5: 7–23.

Pally, R. (2001). A Primary Role for Nonverbal Communication in Psychoanalysis. *Psychoanalytic Inquiry* 21: 71–93.

Panksepp, J. (1998). *Affective Neuroscience*. Oxford: Oxford University Press.

Posner, M.I. & Rothbart, M.K. (1998). Attention, self-regulation and consciousness. *Phil. Trans. R. Soc. Lond. B* 353: 1915–1927.

Sapolsky, R.M. (1996). Why stress is bad for your brain. *Science* 273: 749–750.

Schore, A.N. (1994). *Affect Regulation and the Origin of the Self*. Hillsdale: Lawrence Erlbaum.

Sierra, M. & Berrios, G.E. (1998). Depersonalization: neurobiological perspectives. *Biologic Psychiatry* 44: 898–908.

Silk, K., ed. (1998). *Biology of Personality Disorders. Review of Psychiatry.* Washington: American Psychiatric Press.

Silk, K. (2000). Overview of biologic factors. *Pediatric Clinics of North America* 23: 61–75.

Simeon, D., Guralnik, O., Hazlett, E.A., Spiegel-Cohen, J., Hollander, E. & Buchsbaum, M.S. (2000). Feeling unreal: A PET study of depersonalization disorder. *American Journal of Psychiatry* 157(11): 1782–1788.

Stern, D. (1985). *The Interpersonal World of the Infant.* New York: Basic Books.

Suomi, S.J. (1999). Attachment in rhesus monkeys. In *Handbook of Attachment: Theory, Research, and Clinical Application*, ed. J. Cassidy & P. R. Shaver. New York: Guilford Press, pp. 181–197.

Taylor, G.J. (1992). Psychoanalysis and psychosomatics: A new synthesis. *Journal of the American Academy of Psychoanalysis and Dynamic Psychiatry* 20: 251–275.

Teicher, M.H. (2000). Wounds that won't heal: the neurobiology of child abuse. *Cerebrum* 2: 50–62.

Van der Kolk, B.A. (1993). Biological considerations about emotions, trauma, memory and the brain. In *Human Feelings: Explorations in Affect Development and Meaning*, ed. S.L. Ablon, D. Brown, E.J. Khantzian, & J.E. Mack. Hillsdale: Analytic Press, pp. 221–240.

Winnicott, D. (1956). Primary maternal preoccupation. *The Child and the Family.* London: Tavistock 1957.

Yehuda, R. (2001). Biology of posttraumatic stress disorder. *Journal of Clinical Psychiatry* 62 Supplement 17: 41–46.

IX. SYNAPSES, CYTOKINES (CK) AND LONG-TERM MEMORY (LTM): AN INTERDISCIPLINARY LOOK AT HOW PSYCHOANALYSIS MIGHT ACTIVATE LEARNING VIA ITS EFFECTS ON EMOTIONAL ATTENTION[1]

FRED M. LEVIN

> "Weeds grow and animals slip about in the night where no man dares to hunt them down."
>
> Loren Eiseley (1971). *The Night Country*, p. 3

I begin with a quote from Loren Eiseley about the darker side of mental life because despite all of our increasing knowledge of mind/brain we cannot completely mitigate human suffering. Obviously, a lot depends upon how any new knowledge we acquire would be applied. My hope is that our latest neuropsychoanalytic insights will lead to the improved application of psychoanalysis clinically, especially to matters of the human heart, which Eiseley wrote about so movingly.

Eiseley was an only child, the son of a poor deaf mother during an age when deaf people were ostracized as "deaf and dumb". He lost his father early in his life, was picked upon frequently by playmates, eventually leaving home in his youth to ride freight trains across the U.S.A. during the Great Depression. His vagabond life lasted until he found an island of hope in the night country of his mind. Against all odds, Eiseley ultimately became the distinguished Chair of Anthropology at the University of Pennsylvania, and one of America's great writers![2] He had learned to collaborate with feared others, but he never forgot the inner darkness.

From time to time in what follows I intersperse the dry technical details on neural networks, memory and learning with a completely imagi-

---

[1] The author wishes to thank Jean Carney, Nancy Desert, John Gedo, Eric Gillett, Masao Ito, Thomas Launey, Stanley Palombo, Michael Posner, Arnold Wilson, and Kasuhiko Yamaguchi for their useful suggestions, while he personally takes full responsibility for the final product, and inevitable mistakes.

[2] On the dust jacket of Eiseley's (1971) book, Ray Bradbury sees Eiseley's writings as likely to "linger for the next century".

nary psychoanalytic interchange between myself and Loren Eiseley![3] This should remind us of the human side of my story.

Psychoanalysis, psychology, and neuroscience each contribute to learning theory. My paper discusses some of the details that make this so, and aims towards a unified theory of learning, but one based on interdisciplinary studies which I believe are significantly changing psychoanalytic theory and practice, and impacting as well on neuroscience and psychology.

There will be three parts to my paper: A summary of the Neuroimmune Network (NIN), including the subjects of cytokines (CK) and synapses; an account of long-term memory (LTM); and a postscript on interdisciplinary research.

The key to Horst Ibelgauft's webpage (1999) on the NIN (Levin 2002) and to this paper is the role of cytokines on synaptic plasticity, via their ability to activate and inactivate genes. To quote Abel, Martin, Bartsch and Kandel (1998): *"The neuronal activation of genes means simply that the molecular basis of memory may hinge on the balance between chemical factors that establish versus those which inhibit synaptic plasticity*" (Abel, Martin, Bartsch & Kandel 1998, p. 338).

As I indicate in the abstract, the novel idea of this paper is that *when, under the influence of a psychoanalysis, new opportunities for learning occur, this can reasonably be ascribed to the activation and creation of novel synapses, produced simultaneously with the psychoanalytic maneuvers. First the new synapses help form novel neural networks in the brain of that analytic patient; later these networks will lead to new LTM, that is, new learning*. In what follows I shall attempt to eliminate doubt about the truth of these assertions.

For a long time the Holy Grail of neuroscience has been understanding the cellular and molecular chemistry of memory and learning, particularly the synaptic matrix in which CK act. I believe that psychoanalysts can participate in this quest, along with brain scientists, through the search for apt mind/brain correspondences (Solms 1999). Such correspondences are required for the creation of comprehensive evidence-based learning theories that cross disciplinary boundaries. Therefore, in the second part of this paper I briefly revisit my earlier research into

[3] The reader should not confuse my speculations about the imaginary Eiseley, later in this text, with the real man I introduced on page 1. As noted, they are merely imaginative speculations to illustrate various points about psychoanalysis, and any similarity to the real Eiseley is purely coincidental.

how psychoanalysis facilitates learning readiness (Levin 1991, 1997, 2003). Because our theories and our techniques are inseparable we owe it to our patients to develop both as carefully and objectively as possible.[4]

Learning is an emotional phenomenon, that is, a highly motivated action which satisfies vital needs, and which has its affective origins deep within the brain; it is not a receptive passivity. *That learning involves a core emotional self is particularly important because there is reason to believe that modern cognitive neuroscience approaches to learning are seriously deficient* in two ways: 1. they tend to downplay emotions because feeling states can be hard to qualify and quantify experimentally, and 2. they often restrict attention to those brain structures which can be seen on scans, which means that deep structures such as the periaqueductal gray (PAG) and the hypothalamus within the brain's core tend to get short shrift. Because of the dominance of cognitive theories of learning and memory we need to learn more about how emotions contribute to the sense of self and related learning. Obviously, emotions *and* their substrates deserve our attention.

The biology of learning and memory involves various levels of brain organization. A major problem in discussing brain is that going deeply into such subjects as synapses, ion channels, chemical signaling, receptors, second messengers, etc. is very likely to lose the attention of many people. Yet we cannot avoid such considerations if our goal is to set the stage for a serious neuropsychoanalytic consideration of learning and memory, and particularly if we wish to comprehend the complex role of CK.

From a cognitive perspective memory comes in at least two flavors (explicit and implicit), each further dividing into short vs. long-term forms. Explicit memory involves the medial temporal lobe, (including of course the hippocampuses and related structures), while implicit memory involves the cerebellum, basal ganglia, and the amygdalas. In parts I. and II. I will attempt to review the neuroscientific basis for learning and memory.

---

[4] If one asks how it matters for our understanding of psychodynamics that the mechanisms discussed in this paper and not others are operating, the answer is as follows: reliable treatments, like reliable prediction in any science, rest on accurate understandings, and these can only be discovered by clinical and laboratory experimentation. Proper scientific technique and related theory are based upon various kinds of hard evidence, and many generations of such observations and correlations will be required for real progress to occur.

Psychoanalytically-oriented research in the area of learning and memory has benefited greatly from the work of large number of scholars, a sample of which I wish to mention now: the late Niels Lassen (on working memory), Michael Posner and Tim Shallice (on executive control), Jaak Panksepp (on core affective systems), John Gedo (on the hierarchical-developmental model), Arnold Modell (on the holding environment), Howard Shevrin (on subliminal perception), Colwyn Trevarthen (on mother-infant interaction), Merton Gill, Karl Pribram, Karen Kaplan-Solms and Mark Solms (on mind-brain correlations), and Eric Kandel and Masao Ito (on synaptic mechanisms, including those of the cerebellum) (Levin 2003; also see Panksepp 2000, 2001).

Standing on the shoulders of these scholars and many others, a picture is slowly emerging regarding how the different levels of mind/brain organization are integrated. For example, we know now that during psychoanalysis *learning-readiness is created by the analyst's provision of a proper holding environment, especially as this invites the anlysand's sense of safety and spontaneity. But as Fonagy (1998) has pointed out, what gets learned within any psychoanalysis is more than the consequence of interpretations per se;*[5] *analytic learning obviously involves conditioning (procedural learning) as well, so that in the final result, both emotion and insight play critical roles. And if learning is involved, obviously, clearly changes in synaptic plasticity must be stimulated by the analytic method.* Or, putting it differently, it seems likely that the learning known to occur during analysis has a basis in mind and brain, and this is likely to involve changes in the plasticity of neurons working within a neural network.

In referring to spontaneity I am considering two aspects of psychoanalysis: 1. the patient's free associations in the context of a trusting (i.e. safe) relationship with his or her analyst, and 2. his or her spontaneous transference enactments. What is important is that *it seems to be the spontaneity that activates working memory for those subjects that are remembered or enacted in treatment* (Levin 1997). *Moreover, when these memories and enactments are interpreted to the patient they begin to effect the patient's response patterns, and ultimately enter emotional attention more pervasively.* I am describing a process of learning via emotional attention. This learning occurs whether or not the important

---

[5] This is not meant to depreciate interpretations within analysis, but rather to elevate the role of analytic associative learning (conditioning) that can profoundly alter mind/brain organization and personal adaptive functioning.

analytically-observed behavioral or affective patterns ever actually enter the patient's or analyst's consciousness: this follows from what we know about subliminal perception, and its result is knowledge without awareness (see Levin 1997, 2003 for review of Shevrin's work on this subject).

But even more important, *the emotional attention and the learning stimulated within a psychoanalysis obviously have a direct relationship to the neurochemical changes occurring within the brain during such states*. I will be arguing that the psychological and the neurochemical changes associated with learning are dependent concomitants. I believe that *what Freud called "the drive organization of memory" was his intuitively correct depiction of an adaptive learning process*[6] *in which synaptic plasticity is altered by affective experience (i.e. needs)*. This latter viewpoint, however, alters psychoanalytic thinking in a revolutionary way.

I use the expression *emotional attention* for precise reasons. Emotional because I feel that *what first appears as a novel neural learning network, ultimately becomes new long-term memory (LTM). And the novel neural network starts from our being moved by the emotional significance of particular events, as well as from our multiple emotional needs*, e.g. for companionship, affection, novelty, and so forth. It is likely that our emotionality is what becomes the trigger for new learning, and for the mind/brain changes that support it (see below). And regarding the phrase emotional attention I use the word attention rather than the word consciousness because, as I noted, many important observations enter into our thinking subliminally.

I shall report in this paper upon details of the chemical and neurophysiological changes associated with neural network formation. Our ultimate goal, of course, is to connect, as much as possible, psychoanalytic learning and those factors contributing to new LTM (Levin 1991, 1995, 1997, 2002). My intuition is that *the key to understanding learning within psycho-analysis lies in linking two approaches: 1. those practical procedures within psychoanalysis that invite the activation of working memory and emotional attention, and 2. the psychology of emotional attention and our growing understanding of how insights relate to changes in synaptic plasticity chemically, especially via their effects on CK-mediated gene activation or its matrix.*

---

[6] Thanks to John Gedo for this observation (personal communication).

Finally, in the third part of this paper I make some general remarks about the current profusion of schools within psychoanalysis. In my opinion this reflects interdisciplinary and collaborative efforts which are currently informing psychoanalysis with the viewpoints of various specialized areas of knowledge; but, in addition, *they are also helping establish several new fields, and not just schools*. For example, in the case of efforts to bridge psychoanalysis and neuroscience, they have created neuro-psychoanalysis. This represents a critical paradigm shift in our field, not merely an adjustment of psychoanalytic theory.

1. THE NEUROIMMUNE NETWORK AND IMMUNITY

Horst Ibelgauft (1999) at the University of Munich has created a *C*ytokines *O*nline *P*athfinder *E*ncyclopedia (COPE), containing over fifteen thousand scientific references and 50,000 hyperlinks to the thousands of CK listed. He makes clear that the central nervous system (CNS) and the immune system (IS) (including the neuroendocrine system) essentially constitute a single, integrated overarching system, the *neuroimmune network* (NIN). His conclusion is based upon evidence such as the following: 1. "Neural targets controlling thermogenesis, behavior, sleep, and mood are affected by the so-called pro-inflammatory CK [namely, interleukin 1 (IL-1), interleukin 6 (IL-6), and tumor necrosis factor alfa (TNFalpha)] which are released by macrophages and monocytes during infection"; 2. within the CNS "production of CK has been detected as a result of brain injury, during viral and bacterial infections, and in neurodegenerative processes." In other words, communication between the nervous system and the soma is two way (Ibelgauft, cited in Levin 2002). Moreover, 3. research on the details of synaptic transmission confirms Ibelgauft's conclusion CK play a critical role (for details see Abel, Martin, Bartsch & Kandel 1998); and 4. there are structural chemical similarities between the CNS and IS, for example, Agrin, an aggregating protein, is expressed in lymphocytes (where it plays a role in T cell signaling and lipid raft formation, the latter subject mentioned below), as well as in neuronal signaling, and even in the functionality of the myoneural junction (Khan, Bose, Yam, Solski & Rupp 2001).

So what are CK? They are "a diverse group of soluble proteins and peptides which act as humoral regulators at nano- to picomolar concentrations... and which act on a wider spectrum of target cells than do hormones" (Ibelgauft 1999, section on cytokines, p. 2). "CK [are produced by] cells of the immune system, by glial cells, neurons, and by skin cells" (Levin 2002). You will recall that skin cells and neurons begin as neuroectoderm, that is, they are embryologically related to each other.

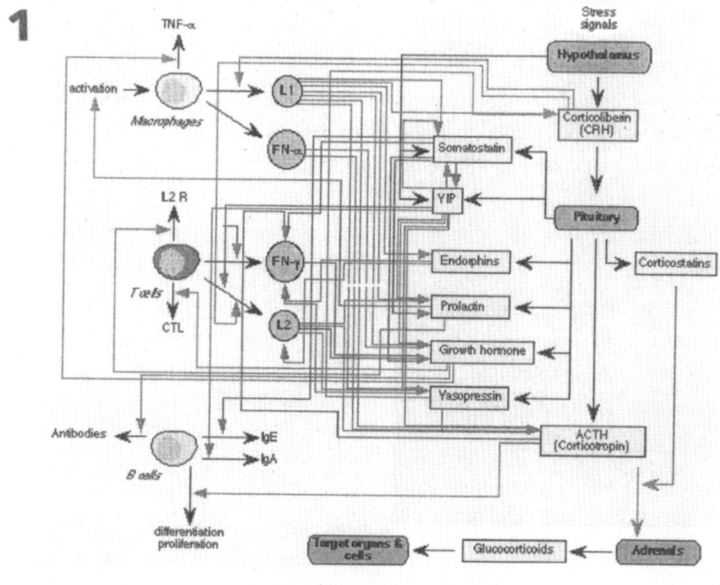

Figure 1

Figure 1 (from Ibelgauft 1999) presents a simplified scheme of the hypothalamic-pituitary axis (HPA).

According to Maier and Watkins (1998), there are at least three pathways between the periphery and the brain. Unidirectional pathways include Pathways A and B. Pathway A is the autonomic nervous system, the sympathetic part of which enervates immune organs. In Pathway A, catecholamines (largely epinephrine and norepinephrine [NE]) are released from sympathetic nerves controlling various organs (e.g. the spleen, thymus, etc.).

Pathway B is the hypothalamic-pituitary axis (HPA). In Pathway B various releasing factors (e.g. corticotropin releasing factor [CRF] whose source is the hypothalamus) enter into the bloodstream heading from hypothalamus to pituitary, which in response releases various hormones on demand (e.g. adreno-corticotropin releasing hormone, or

ACTH). These pituitary hormones then travel via the blood stream and effect both the peripheral endocrine organs (e.g. the adrenal cortex, which then releases cortisol) and the immune system.

I will jump to the bidirectional immune pathways between the brain and the periphery. One arm of Pathway C is composed of *immune cells* (i.e. macrophages and other cells) that releases IL-1 (as an example) which *can stimulate sensory paraganglia of the vagus nerve, which in turn carries impulses to the nucleus tractus solitarius (NTS) and the area postrema (AP) in the brain-stem.* We do not often think of the vagus nerve as having a sensory component but it does. A neural cascade then occurs which at least theoretically could result in the release into the blood of CK such as IL-1. There is some debate, however, whether any of this blood born IL-1 actually reaches the brain itself or crosses the relatively impervious blood-brain barrier. Thus, possibly *more important for CNS CK production is a second arm of pathway C which starts as does the first in the vagus nerve, reaches the NTS and AP, but then heads into two different brain areas, the hippocampus and the hypothalamus, wherein IL-1 is generated directly within the brain* (Levin 2002).

Once so generated in the brain, CK (or the IL 1 in our example) are capable of interacting with any neuron that carries the appropriate CK receptors. Any of three further different reactions then become possible. One is simple neurotransmission, i.e. the generation of a postsynaptic depolarization. Second, there could be a response specific to the CK involved; for example, in the case of IL-1, there would be a proinflammatory response, namely sickness reactions, fever, fatigue, anorexia, etc.

The third possibility is the induction of gene transcription and translation (see below),[7] *which obviously can profoundly effect memory and learning* (Levin 2002). CK are thus part of the mechanism for LTM formation, which I intend to review in Part II.

[7] The precise meanings of the technical terms "transcription", "translation", and "copying" within genetics are not obvious, but may be distinguished from each other as follows: transcription refers to the copying of DNA into messenger RNA (mRNA); translation refers to copying RNA code into related proteins (which then express gene functions); incidentally, copying refers to the creation of new forms of RNA from other RNA (for example, under the influence of RNA-dependent RNA polymerase) (Ahlquist 2002, p. 1270).

1. 1. KINDS OF IMMUNITY

Some will appreciate a brief review of immunity to place the subsequent discussion of CK research in a more familiar context. In a recent review of the subject (Levin 2002) I cite Maier and Watkin's (1998) description of immunity as follows: "There are two kinds of immunity, either specific or nonspecific. The former is also called 'humoral' and is slow; the latter is also called 'cellular' and is relatively rapid. Both responses are secondary to the detection of foreign invaders, either viral, bacterial, or toxic-chemical, but they follow distinctly different pathways."

Nonspecific cellular immunity occurs within 1–2 hours of the invasion, and depends upon phagocytes such as macrophages or neutrophils. Later, when I summarize specific immunity, I will note the role of macrophages in that process as well, where they serve as antigen presenting cells (APCs); however, in the present discussion of rapid, nonspecific immune response the goal is three fold: 1. the nonspecific recognition of foreign material (antigen, none self, "stressor", "toxin"), 2. the transfer of this material into phagocytes ("eating cells"), and 3. its destruction by various intracellular means, including exposure to enzymes, nitric oxide, and/or various CK (e.g. IL-1, IL-6, TNFalpha, etc.). These defensive proteins and other transmitters create the so-called inflammatory reaction which further attracts other immune cells.

Specific humoral immunity is more complex, generates antibodies, and takes 3–5 days because a number of cell reproductive cycles are required to create enough of those cells making the specific antibody targeted at particular antigenic stimuli. First the foreign substance (antigen) is engulfed within the APCs. Parts of the invader are then moved to the surface of the APC. This display of parts is recognized by T cells which carry appropriate surface receptor sites. Recognition involves a binding between the T cell receptor site and the displayed antigenic part, a process that stimulates the T cell to differentiate and proliferate, increasing the number of T cells which can then target the specific antigen (invader).

Some T cells are cytotoxic, and kill specific antigen producing cells. Others are T-helper cells, which help a separate class of B cells expand their number, and begin to detect antigen, differentiate, and proliferate just as the original T cells did, thus joining the battle against antigen. From our point of view, what is important is that the substances secreted by the APC which assist their performing complex antigen site matching and intercellular communication among the T-helper cells, and the B cells, are basically CK (e.g. the interleukins already mentioned).

1. 2. CK AND SYNAPTIC PLASTICITY

Since CK are involved in gene activation it will help to define the processes of gene transcription and translation. Transcription involves forming messenger RNA (mRNA) from DNA, a process mediated by specific RNA polymerases active at the ribosomes (Ibelgauft 1999, section on Gene expression, p. 1). The second step, translation, involves mRNA being translated into specific proteins which express functions for particular genes.

As we shall see, many complex factors effect gene expression in addition to the CK I am highlighting in this paper (Levin 2002). For example, all genes have *promotor regions* which contain special DNA that regulates transcription and processing of the genetic codes. Transcription factors which aggregate in the promotor regions[8] are referred to as *response elements*. Some of these transcription factors are required for the transcription of all genes, but others are important primarily or exclusively for special classes of genes. One subgroup of transcription factors or so-called nuclear receptors is called *early response genes* (ERGs). We will return to ERGs later when we consider the cascade leading to LTM, but first I need to describe the synapse and types of memory.

1. 3. THE SYNAPSE[9] AND PRIMING OF PLASTICTY

Everyone knows that synapses are the spaces or clefts into which neurotransmitters are released, thus providing chemical contact between

---

[8] Within the promotor region are docking components where the many transcription factors attach to each other and the DNA, which enable RNA polymerase to align itself next to the gene which is to be replicated.

[9] One complication of the learning picture presented is that although previously thought impossible, it is now known that new neurons can appear in the brain in various locations. Primate research has shown, for example, that prefrontal, inferior temporal, and posterior parietal cortex is important for cognitive functions, and these three zones demonstrate new neurons that originate in the subventricular zone and migrate through the white matter to the neocortex, where they extend axons. This continues throughout adulthood (see Gould, Reeves, Graziano & Gross 1999). Unfortunately, this aspect of learning cannot be covered in this paper because of space and knowledge limitations.

neurons. These transmitters are released from presynaptic vesicles, traverse the synaptic cleft and attach themselves to receptor sites on the postsynaptic membrane (that is, the membrane of the second neuron). This usually allows the first neuron to depolarize the second neuron and continue a nervous excitation wave in a particular direction.[10] There may be as many as a million million synapses within the CNS. What may not be readily appreciated, however, is that the actual structure and functional relationships of synapses belie the simple description just given! For one, *synapses do not just preexist, but are creatable on emotional demand by the activity of cytokines, that is, in relation to the drive towards learning*. Second, there is evidence that several proteins are complex regulators of presynaptic plasticity, effecting the three basic steps of neurotransmitter relase, as follows: Rab3A and RIM appear to be involved in vesicle docking, Munc 13 is required for priming, and synaptotagmin effects vesicle fusion (Dobrunz & Garner 2002). As these authors put it, Schoch et al. (2002) and Castillo et al. (2002) have shown that "the presynaptic 'bouton' is filled with synaptic vesicles loaded with neurotransmitters. After a priming step these vesicles dock at a specialized region of the presynaptic plasma membrane known as the active zone. Electrical impulses arriving at the active zone trigger the fusion of the synaptic vesicles with the plasma membrane, releasing their neurotransmitter into the synaptic cleft" (Dobrunz & Garner 2002, p. 277).

Let me take a brief pause. If I imagine Loren Eiseley as my patient in psycho-analytic treatment, I believe the critical issue for him would be whether or not I would be carefully listening to him and remembering what he has told me. Only then would he be reassured that he exists for me, i.e. in an important emotional sense. My hunch is that this kind

[10] Another major complication of this simplified picture is that evidence is accumulating that there are also retrograde messengers informing the presynaptic membrane of events in the postsynaptic membrane, controlling presynaptic membrane neurotransmitter vesicle formation and release. Nitric oxide may play a role here, and arachidonic acid (Bliss 1988, p. 88, see also Linden & Routtenberg 1989, especially Fig. 2, p. 289). In addition, endogenous cannabinoids can function as retrograde messengers activating CB1 receptors on the presynaptic side, suppressing neurotransmitter release. CB1 receptors are "one of the most abundant ... in the brain" expressed at high levels in the hippocampus, cortex, cerebellum, and basal ganglia, accounting for the striking effects of cannabinoids on memory and cognition (Wilson & Nicoll 2002, p. 678). Also, another retrograde signaling pathway involves EphB receptor-Ephrin B ligand interaction in hippocampal mossy fiber-CA3 synapses (Contractor et al. 2002).

of emotional awareness or attention is what serves as a decisive trigger for the induction of new synapses and the use of these in the service of new learning. Whether we are seen, heard, appreciated, and especially loved and respected, is never neutral.

Emotional attention in psychoanalysis may be particularly useful because it feels to us as if it is not guided, that is, it is something felt to be independent of outside agency and thus of immediate interest to the self. In spontaneous activity emotional attention and executive functions seem less in play than they would be otherwise; and perhaps that is why this kind of attention feels important, that is, it seems at time particularly salient for our survival. Selective attention may therefore be a precursor to emotional attention. Csikszentmihalyi argued long ago (1975), and I agree, that emotional attention plays a central role in learning, especially when it falls in some optimal effective range.[11]

Returning to synapses, it is obvious as we make our way through the very dense subject of what happens in the synaptic cleft, as well as at the pre- and postsynaptic membrane levels, that there are many factors which influence synaptic weights and the related learning in networks of neurons. Exploring this complicatedness of synapses is important, however, if we are to understand CK activity within the context of neural transmission, memory and emotional learning. It will therefore help to consider the categories of memory and say something as well about receptors.[12]

As I noted in my introduction, it appears unanimous that there are at least the two major types of memory, implicit and explicit (sometimes called procedural and declarative), at least two kinds of receptors (AMPA or NMDA receptors[13]), and the memory types divide neatly anatomically (for explicit memory – lateral temporal lobe and hippocampal vs. for implicit memory – the cerebellum, basal ganglia, and amygdala's). The time course of activity of these two memory systems is also different (in the case of explicit memory, multiple repetitions seem required for learning, and memories are read out by merely asking

[11] Jean Carney, personal communication.

[12] Squire divides memory into declarative (explicit) and non-declarative (implicit); the latter also being called procedural. Explicit involves memory for facts and events; implicit involves memory for skills, priming, simple classical conditioning, and non-associative learning (Squire 1998, p. 62).

[13] NMDA and AMPA receptors are two glutamate receptor families, glutamate (glutamic acid) being the major excitatory neurotransmitter in the brain (Bliss 1998, p. 88-89).

conscious questions. Nonconscious or unconscious methods can also result in or tap explicit memory. In the case of implicit memory, however, learning usually occurs with a smaller number of passes,[14] and recall usually obtained by priming, i.e. without consciousness being required). Children do as well as adults in the case of implicit learning.

Another time out. My hypothetical treatment with Eiseley would involve both explicit and implicit memory systems. The implicit system would especially be involved whenever I would prime his memory by reminding him of things he told me earlier, and which might be related to his here and now feelings. For example, imagine that he was talking in the previous session about people who interrupt him with stupid comments instead of patiently listening to him. Suppose I ask him, towards the end of his spontaneous comments, if when he was complaining about bad listeners last time he was also thinking that I fit into this category. This could be seen both as an interpretation, but also as a case of my priming his memory. Another kind of priming, might be to let him know what I think he seems to be feeling, and ask if this might be the case or not. Or priming could be my mentioning a song that is running through my mind, and either sing it to him or wonder aloud with him if it reflects something in him that I might be tuning in upon. Priming connects past and present feelings in complex ways.[15] Priming is also the critical mutual activity which connects the minds and brains of two individuals, that is, which puts them in resonance with each other.

Long-term-potentiation (LTP) has been eloquently reviewed by Bliss (1998, pp. 78–82; see also Johnston 1997 and Malenka & Nicoll 1999), *and is likely the precursor of LTM* (Bliss 1998).[16] LTP is defined as the

---

[14] Like any generalization, my comment about the number of passes required for new implicit memory formation is only relative to explicit memories. Implicit memory formation also requires repetition. Moreover, given our limited present knowledge, it is probably wise not to insist that there are only two major types of memory without overlap, since implicit memory is likely a property of all or at least many neural networks. Recent studies of perceptual learning have further shown clear evidence of V1 modification as a result of training, and it would be surprising that such modification would not occur in many other networks (Michael Posner, personal communication).

[15] For additional information on priming see Schacter & Buckner 1998.

[16] Bliss states that arguing that LTP is not the basis for LTM brings to mind Mark Twain's remark about those who assert that Homer may not have been the author of the Iliad: "If it wasn't Homer, then it must have been another old blind Greek poet of the same name [!]" (Bliss 1998, p. 91). Incidentally, LTP could also be seen as the variable that relates LTM and neural networks to each other, as that

easier neural firing secondary to neuronal use or stimulation. LTP is also characterized by so-called specificity and associativity. These are defined as follows:

In terms of the hippocampus, for example, if two sets of converging electrodes are placed on either side of a recording electrode in the CA1 region of the hippocampus, say to activate pyramidal cell axons of the neighboring CA3 region, LTP will be produced in either of the two converging pathways (say CA1 and CA3) providing only that it is active at the time of the stimulating input. In other words, each synapse, in principle acts as an "independent computing device" (this is *input sensitivity*; Bliss 1998, p. 80). *Associativity* refers to the ability to induce LTP in the second pathway of the above example, that is, even in the face of a weak tetanus in this pathway, when a strong stimulus is given at the same time to the first pathway. Bliss sees this form of associative memory as coinciding with classical conditioning (Bliss 1998, p. 81).[17]

Each form of memory, implicit or explicit, is also recordable in two phases: short-term and long-term. However, *only long-term memory requires the synthesis of new proteins* (Albright, Jessell, Kandel & Posner 2000, p. S13). "Short-term synaptic enhancement [...] results from a covalent modification of preexisting proteins mediated by the cAMP-dependent protein kinase A (PKA) and by protein kinase C (PKC)" (Albright et al. 2000, p. S13). In contrast, *long-term memory involves "the translocation of PKA and mitogen-activating protein kinase (MAPK) to the nucleus of the sensory neurons where these kinases activate CREB1 and derepress CREB2, leading to the induction of a set of immediate response genes [ERG] and ultimately resulting in the growth of new synaptic connections"* [emphasis added] (Bartsch et al. 1995, 1998, cited in Albright et al. 2000, p. S13).[18] The point here is that at a

---

activity which leads to "the next step". In other words, other things being equal, LTP leads to neural network formation leads to LTM, more or less.

[17] A further complication touched upon is that something needs to turn off LTP once it has been induced (Bliss 1998, p. 89). This is one aspect of long-term depression (LTD) described most extensively by Ito (2001).

[18] It would seem that the preceding paragraph accurately summarizes a lot of what is known about synapses, at least so far as signal transduction leading towards long-term memory is concerned. However, this is of course an illusion. The real complexity, for example, also would need to include many other subjects, such as lipid rafts. Sphingolipid- and cholesterol-rich domains in the plasma membrane of cells are known as lipid rafts. Within this subcellular domain or compartment proteins (cytokines) are sequestered, and this "increases the specificity and efficiency of signal transduction" (Zacharias, Violin, Newton & Tsien 2002, p. 913). Also, within

molecular level all of the known memory systems likely involve roughly the same set of molecular [cytokine] markers (Michael Posner, personal communication).[19]

Returning to the implicit/explicit memory distinction, it is important that in spite of the differences noted, the explicit and implicit memory systems should not be thought of as wholly separate and competitive; rather, they are judged to be highly complementary to each other in the brain and usually create a mosaic of integrated memory formats, i.e. "they ordinarily cooperate with each other in learning" (Squire 1998, p. 68). This is part of the complexity of memory alluded to.

Putting this all together in 1985 Floyd Bloom described the neural cascade, culminating in LTM, as follows: *it begins when NE alters the firing of particular neurons by activating cyclic AMP* (an intracellular second messenger), which then activates the enzyme protein kinase (noted above), which leads to the phosphorylation of macromolecular substrates (such as pre- and postsynaptic ion channels) and the enhancement of transmission. Obviously, by the mid 1980's, synapses were understood to be fairly complex structures even though the wrinkle of gene activation and CK was less well understood compared to now, and much remained to be discovered.

Contemporary articles on synaptic activity clearly show the growing appreciation and complexity of CK in synaptic changes (and make clear the large number of other cofactors involved – see below on Ito's work). I will mention two in particular. The first is by an American research group, Beattie et al. (2002) which I introduce to make clear the role of glial cells; the second paper, characterizing cerebellar long-term depression (LTD) cerebellum, by Masao Ito (2001) is introduced to explain

these spaces protein-protein interactions "maintain many signaling complexes" (Zacharias et al. 2002, p. 913). We are only gradually learning about the nature of lipid rafts and their controls.

[19] The time course of neurotransmitter (cytokine) induced activation of genes makes clear what leads to what: first neurotransmitter binds to receptor, very quickly (within a few minutes) this is followed by the curve of cAMP response, followed by the curve representing the activation of Protein Kinase, then activation of CREB, the activation of ERG and the protein synthesis it leads to (e.g. Fos or Jun) in sequence (Stahl 2000, p. 8, see especially Fig. 2). Curves for late-gene mRNA synthesis, late-gene protein synthesis, and long-term effects of late-gene products all follow the previous cascade, in order, as one would expect (i.e. the effects of early genes produce products which can activate late genes starting within 60 minutes or so).

LTD and critical synaptic subtleties which might otherwise go unnoticed.

The American group describes how "activity-dependent modulation of synaptic efficacy in the brain contributes to neural circuit development and experience-dependent plasticity." They further remark on the decisive role of glial cells as follows: "although glia [supporting cells] are affected by activity and ensheathe synapses, their influence on synaptic strength has largely been ignored" (Beattie et al. 2002, p. 2283).[20] The key is that "a protein produced by glia, tumor necrosis factor alpha (TNFalpha) [a CK] enhances synaptic efficacy by increasing surface expression of AMPA receptors" (Beattie et al. 2002, p. 2284) thus playing a role in synaptic plasticity.

To be more precise, however, in spite of explicating the role of glial TNFalpha in activating genes and thus altering AMPA receptors, these researchers did not conclude, at least in their particular set of experiments, that gene activation played a major role in modifying synaptic plasticity; rather, they felt that activity was responsible for the changes in AMPA receptors. Nevertheless, they believe gene activation could not be ruled out absolutely (personal communication).[21] Obviously, it may not be easy to tell when gene activation is a direct factor in synaptic enhancement, or when indirect. The distinction may not be that critical to our synopsis of learning changes, however.

The complexity of the synapse, as regards CK, emerges still more clearly if we shift to Masao Ito's work (2001). Ito is Director of the Brain Science Institute (BSI) of RIKEN[22] FRONTIER INSTITUTE in

---

[20] It has been known that glial cells increase the ability of neurons to form and maintain synapses, by as much as a factor of 7, but prior to this study, it was not clear how this effect was obtained (Helmuth 2001). A second synapse-promoting signal is apparently the release of cholesterol from glial cells (the work of Mauch et al. 2001). Mauch et al. in turn credit Pfrieger and Barres for their work in 1997 on retinal ganglion cells (Pfrieger & Barres 1997).

[21] M. Beattie (personal communication) states that he agrees that the effects of TNF alpha on gene expression are very interesting and could be important in the long term regulation of glutamate receptor sensitivity. In the work reported, though, the effects on AMPA receptor surface localization take place within minutes. The previous data from the Malenk/Stanford group and others suggest that there is considerable regulation by modulation of AMPA receptor recycling without (necessarily) the induction of changes in gene expression. Whether longer term effects may be mediated by transcriptional/translational effects remains to be seen."

[22] RIKEN is an abbreviation of "rikagaku kenkyu jo" which in Japanese is Institute for Chemical and Physical [Physics] Research.

Wako, Japan. If you examine Fig. 2 from Ito's paper LTD, I think you can begin to see the complexity I am talking about. It seems that synaptic phenomena are not only complex, they are probably more complex than most of us can even imagine!

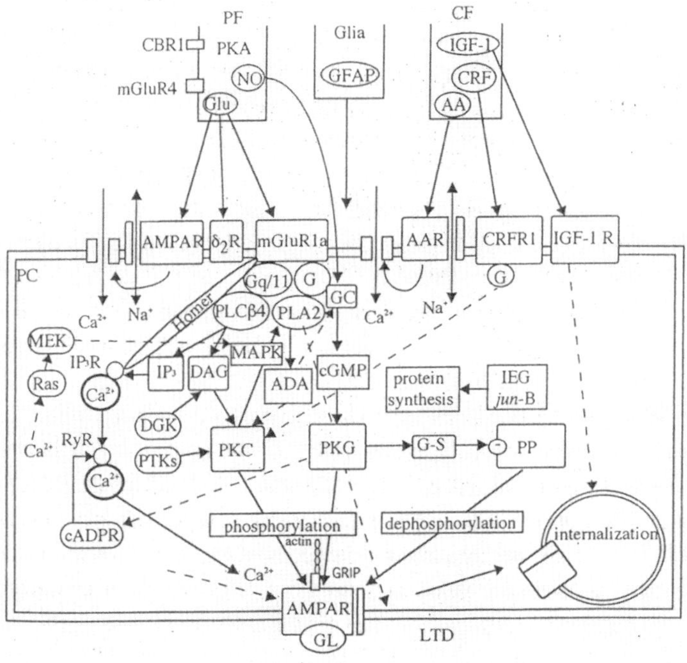

Figure 2

Let me explain some of Fig. 2. It will help to consider the following points:

(1) Ito was early to demonstrate LTD, the experience-based down regulation of neuronal firing (i.e. the opposite of LTP). The required trigger for LTD in the cerebellum is when Purkinje cells are simultaneously fired off (conjunctive stimulation) by numbers of granule cell axons (GA, also abbreviated as PF for parallel fibers), and by single climbing fibers (CF). It is then that the AMPA receptors (AMPAR) of Purkinje cells demonstrate LTD induction. This is illustrated in Fig. 2.

(2) The GA carry the stimuli from basket cells (BC) which carry input from two potential sources, either spinal cord and brain stem, or from the cerebellar flocculus. The CF carry input from the inferior olive or the retina.

(3) When the input is from the flocculus and retina, the cerebellum is able to provide the vestibular occular reflex (VOR), which compensates for head movements by eye movements to keep retinal targets stable. Ito has done much to understand the basis for the VOR.

(4) In Fig. 2 the upper part of the diagram shows the simultaneous input from PF (i.e. GA), and CF, along with glial input. At the bottom the rectangle stands for the PC, with its AMPAR. Within the rectangle Ito diagrams all the subprocesses, namely (a) the initial processes of signal transduction (including the now familiar first messengers, receptors, ion channels, G proteins, and phospholipases), (b) processes of the transduction itself (including ion concentrations, second messengers, protein kinases and protein phosphotases, activation of the AMPA receptor, and the chemical network for LTD), and, (c) the roles of various other complex factors.

(5) It is important that unlike LTP, in the case of LTD "there is no evidence indicating that adenyl cyclase, cAMP, or PKA has any role in [LTD] induction" (Ito 2001, p. 1155).

At this point I think you can well understand how my initial assertion that CK effect memory and learning, stated as it was without immediate further qualification, came dangerously close to inviting a grossly oversimplified view of the synaptic changes behind memory and learning. Obviously, on the one hand, one needs to be very careful in making a judgment about the nature and degree of CK effects, that is, CK do not operate in isolation. And yet on the other hand, CK-induced gene activation clearly plays a decisive role in synaptic plasticity, especially as regards the establishment of LTM.[23] What remains for neuroscience to determine is the exact involvement of CK in various types of changes in

[23] In particular, although Ito does not discuss cytokines as a separate topic in his paper, he does include no less than four cytokines within his map of synaptic activity: IGF-I (Insulin-like Growth Factor- I), IEG (immediate early gene/ERG), MAPK (MAP kinases; also called Mitogen-activated protein kinases, or ERK for extracellular signal-regulated kinases), and BDNF (for brain derived neurotrophic factor). Regarding BDNF, it is especially important that recent work shows its ability to provide anterograde stimulation of synapses, something of a surprise (Kohara & Tsumoto 2002). Postsynaptic neurons "have been thought to release neurotrophic factors which promote extension of axons and differentiation or survival of presynaptic neurons" (Kohara & Tsumoto 2002), but this transfer is towards the postsynaptic side and it is clearly activity-dependent. This makes clear how we are needing to adjust our oversimplified picture of synapses to meet the new facts.

synaptic plasticity, and the factors which control these changes, as, for example, Ito has provided par excellence for the cerebellum and LTD.

Before leaving this section let me distill more clearly the significance of Ito's article: As I read it *Ito is postulating that the detailed memory mechanisms which he and his coworkers have elaborated for the cerebellum and motor learning in particular might very well apply equally to all implicit learning.* As he puts it, there are "functional roles of the cerebellum in implicit memory and learning in general" (Ito 2001, p. 1182). His second important conclusion is that the *"investigation of gene regulatory mechanisms for LTD will be of essential importance* [emphasis added]. A possibility [also] exists that LTD as functional depression is consolidated as a structural change in synaptic contacts and spines" although he adds "as yet there is no evidence showing morphological changes correlated with LTD" (Ito 2001, p. 1182). Obviously, it will be important to examine each of the elements in the implicit memory system and see which are actionable via gene activation and which by other means.

One additional comment on Ito: having known him personally for a long time I want to remind you if you don't already know it that Ito was the researcher who realized that the cerebellum moves ideas, not just limbs, and is thus deeply involved with learning generally, and not just motor learning. I have often used this idea of his in my own analytic work, especially in the sense of paying attention to the gracefulness of a patient's movements or lack thereof, thinking this can sometimes provide important clues to what the patient is feeling, and where they might be developmentally, but not otherwise communicating except in a kind of pantomime.

I further imagine, that if Loren Eiseley were in treatment with me he would communicate a lot in such pantomime, and it might well be my place to comment upon it, that is, if I could find a non-intrusive way of doing so. This might conjure up in his mind recollections of his interaction with his deaf mother, who was undoubtedly communicating with him in sign language or via gestures, a very special kind of communication that was once quite embarrassing for deaf people to use in public, and for their hearing children as well, although nowadays this situation has greatly improved as sign language has achieved the status of a formal language, on a par with any other. Importantly, if I used my knowledge of sign language to communicate with Eiseley, at the right time, he might respond to me very differently, that is, with more emotional associations than if the same message were communicated by other means with less personal associations for him. Of course, this would be another clue that memory is affectively organized.

1. 4 AMPA and NMDA receptors

Summarizing to this point, there are no less than six possible molecular mechanisms that are candidates for explaining the stable synaptic changes we associate with LTM (see Lisman & Fallon 1999, reviewing the work of Bhalla & Iyengar 1999, p. 381):

a second messenger switch (in which MAPK activates PKC via PLA2, and where a reverse feedback effect can also occur in which PKC activates MAPK via RAS or RAF);

a kinase switch (in which CAMKII, i.e. CAM-kinase II, autophosphorylates itself once activated by an increase in Calcium ion, a process reversible via phosphatase);

a transcriptional switch (such as our mention of CK altering gene expression or undoing it);

a translational switch (starting from mRNA which would give rise to a protein and then to a second messenger, where the latter would stabilize a translational switch in the synapse);

a structural unit change associated with turnover within the synapse (such as hinted at in Ito's article);

a structural unit change without turnover (an example being the action of agrin within the neuromuscular junction, a subject peripheral to the present discussion).

We need also to remind ourselves that just as is the case for synapses, AMPA receptors, and other receptors as well, are createable on emotional demand. That is, receptors "are synthesized in the endoplasmic reticulum, processed in the Golgi apparatus, and transported to the cell surface in membrane vesicles [...]to dendrites" *as they are needed* (Nakagawa & Sheng 2000, p. 2270). The key to their effectiveness in terms of excitatory synaptic transmission effectively relates to the number of such AMPA receptors that are produced, their rate of production and their exact distribution postsynaptically (Nakagawa & Sheng 2000; Shi 2001, p. 1851). Using recombinant AMPA receptors Shi has been able to cleverly prove that AMPA receptors are in fact delivered to synapses exactly in relation to LTP induction (Shi 2001). As noted earlier, this decisively proves that there must be a method for communication from the postsynaptic membrane to the presynaptic side, otherwise, how could the presynaptic membrane prepare and deliver receptors in such a timely way?

The NMDA receptor is also a key player in synaptic plasticity (Ghosh 2002, p. 449). Ghosh, reviewing the work of Takasu, Dalva, Zigmond and Greenberg (2002) points out that "post-synaptic NMDA Calcium ion flux is itself dependent on Eph receptors. Eph receptors are a large family of receptor tyrosine kinases that regulate various developmental events", a reference to the cytokines EphA and EphB (Ghosh 2002; Takasu et al. 2002).

Takasu et al. summarize their findings in the following way: The EphB receptor tyrosine kinases are localized at the NMDA synapses and their role is to effect activity-dependent and activity-independent development and remodeling of synaptic connections (Takasu et al. 2002, p. 491). The importance of such changes in NMDA synaptic reinforcement is that such modifications in the synapses becomes the basis for changes in the hippocampal CA1 region, which is crucial for converting new memories into LTM in the explicit memory system (Shimizu, Tang, Rampon & Tsien 2000).

2. LTM: FROM EMOTIONAL ATTENTION TO LTM

The combined effects of the CK induced cascade of events results in structural changes in the synapse, in changes within the nuclei of neurons, and ultimately the establishment of LTM which grows from larger numbers of such neurons undergoing related changes and operating as parts of neural networks (i.e. ultimately, as databases of mind/brain).

Regarding the creation of LTM, we can summarize again how this likely comes about, based upon knowledge of those specific brain structures most connected with and affected by emotion and how they are activated. Obviously, CK play a role. I have already described the specific CK involved, and at least some of the factors which stimulate CK activity. And as noted above, one key is emotional attention, that is, the changes (chemical, physiological and psychological) which accompany the activity of neural systems and which provide emotional attention, especially the ventral part of the anterior cingulate cortex (ventACC).[24] *Obviously, the ventACC is a critical component in the system being described.*

---

[24] By emotions I am referring not merely to the cortical systems, and of course the limbic system, but also to the deep structures of the brainstem or core, as pointed out by Panksepp (2000, 2001), including the periaqueductal gray (PAG), hypothalamus, etc.

It will help to briefly review where we have been. As described earlier, LTM "results from the translocation of PKA and mitogen-activated protein kinase (MAPK) to the nucleus of the sensory neurons where these kinases activate CREB1 and derepress CREB2, leading to the induction of a set of immediate response genes and ultimately resulting in the growth of new synaptic connections" (Albright et al. 2000, p. S13, citing the work of Bartsch et al. 1995, 1998).

Ibelgauft, writing under the rubric of CRE (for cyclic AMP-response element), describes the process as follows: "Cyclic AMP-response element cAMP second messenger pathways provide a chief means by which cellular growth, differentiation, and function can be influenced by extracellular signals...an inducible enhancer of genes that can be transcribed in response to the increase in cAMP levels" (Ibelgauft 1999, section of CRE, p. 1).

With the notable exception of LTD and thus most implicit learning (which Ito has clarified does not involve adenyl cyclase), cAMP levels need to be enhanced as part of the basic pathway leading to the growth of new synapses and the changing of synaptic plasticity for LTP. So the question then becomes: What can we say about those factors in explicit memory which elevate cAMP in a timely way, since after all this step within the system is what leads to a cascade ending in immediate response genes (IEG, also called ERG for early response genes) and new synaptic production?

Clearly, cAMP can come from a number of sources. If we consider ATP as the source of cAMP level, then one obvious endocrine initiator, as noted in the early article I cited by Floyd Bloom, would have to be NE[25] which is known, along with epinephrine, both to break down ATP into 3',5'-cAMP and pyrophosphate, and which also is itself released in response to significant emotional stimulation. In addition, as I noted in discussing the immune part of NIN, sympathetic neurons are sources of NE, and we know that NE is of course also a central neurotransmitter.

Another question remaining is have we tied in the affective systems to the NE release, that is, in relation to affective attention neural network formation, and ultimate LTM formation? The answer seems not hard to

[25] This statement is based on the classical work on Hodgkin, Huxley and Katz (1952), and a minor bit of personal research (Levin 1962) demonstrating that the effects of epinephrine on liver essentially are those induced by 3'5'-AMP (cAMP), which epinephrine produces by catalyzing ATP into cAMP and P-P. NE follows this pattern (Bloom, op. cit.).

guess. *The original affective signal leading to NE release would most likely be received by the ventral division of the anterior cingulate, along with the amygdalar, hippocampal (CA1 cells), and hypothalamic cells in response to emotional activity deeper in the brain's core (e.g. the PAG). It may help to remind ourselves that this ventACC "is connected to the amygdala, PAG, nucleus accumbens, hypothalamus, anterior insula, hippocampus, and orbitofrontal cortex, and has outflow to autonomic [sympathetic], visceromotor and endocrine systems"* (Bush, Phan & Posner 2000, p. 216). Once the NE concentration would rise in association with ventACC, and sufficient cAMP would be available thereby (via the breakdown effect on ATP as described earlier), we would expect the cascade of PKA, MAPK, activation of CREB1 and the deactivation of CREB2 (i.e. in response to shifts in cAMP which effect the kinases, phosphorylation mechanisms, and protein protein interactions of great complexity),[26] and finally ERG and its effects.

Regarding immediate response genes (ERG), it should be noted that aside from hormones, cAMP signals and cytokines, ERG is triggered by many factors: antigens, neurotransmitters, leukotrienes, prostaglandins, eicosanoids, protein synthesis inhibitors, calcium signals, cellular adhesion molecules, osmotic pressure changes, and stress, to name but a few (Ibelgauft, op. cit., ERG section, p. 1). From this it should be obvious that many other pathways might also play a role in ERG activation in addition to the ones mentioned (viz. via cAMP and NE effects) and thus effect the final stimulus pathway towards formation of LTM. Whatever the contributing factors, however, clearly the brain is designed to provide LTM on demand by multiple mechanisms such as we have been considering.

From the psychoanalytic perspective, the formation of new neural networks and ultimately LTM in association with learning obviously occurs under a variety of circumstances which relate to emotional and attentive processes, including psychoanalytic interventions.

By the way, I believe Loren Eiseley would have been most interested in this paper, especially in the relationship between the affects and

---

[26] "At least ten different CREB protein genes have been cloned, including CREB, CREB1, CREB2, CREBP1, ATF2, ATF (activating transcription factor), ATF43, ATF1, TREB 36. The CREB1 gene maps to human chromosome 2q32.3-q34. The human CREB2 gene maps to 2q24.1-q32, a site very close to that of the CREB1 gene. CREB proteins are activated by phosphorylation by cAMP-dependent protein kinase A" (Ibelgauft 1999, CRE section, p. 1).

ideas generated in his imaginary psychoanalysis, and their associated specific chemical and anatomical changes in his brain. Remember, as a physical anthropologist he often lived mentally at the intersection between the technical chemical and anatomical findings from human artifacts and the history of human stressful experience.

3. POSTSCRIPT: INTERDISCIPLINARY RESEARCH

I have just presented some mind/brain details connecting the psychoanalytic and the neuroscientific understanding of learning and memory, at least as this involves learning within the psychoanalytic situation and the on demand effects of such on cytokines and synaptic plasticity. In spite of my effort, however, I suggest that it still remains very difficult to align these two perspectives; to date only a small handful of reliable mind/brain correlations and linkages have been generally accepted as valid.[27] Only time will tell about the acceptance of the correspondences I have speculated about today. Where then does hope come from for further bridging between mind and brain?

My answer is that our principal hope comes from collaboration among specialists across disciplinary boundaries, and employing various philosophical approaches; only then do we move bit by bit closer to deeper and specific bridgeable understandings within neuroscience and psychoanalysis.

I noted two influential approaches at the beginning: Pragmatism, and what I called thinking things through deeply and creatively. Regarding the former, Stanley Fish states (cited by Arnold Goldberg 2002, p. 249) that "Pragmatism is the philosophy not of grand ambitions but of little steps; and while it cannot help us to take those steps or tell us what they are, it can offer the reassurance that they are possible and more than occasionally." But we also need to respect the creative search for truths independently of what is practical. This seems to me a complementary area where psychoanalysis excels.

Collaboration across disciplinary, cultural, philosophical and language boundaries represents a much needed effort to reduce the complexity of thinking generally by introducing novel testable scientific

---

[27] In my opinion, credit for the most important mind/brain correlations is particularly due Jaak Panksepp (1998, 2000, 2001), and to Mark Solms and Karen Kaplan Solms, some of which appears in the latter's book (Kaplan-Solms & Solms 2000) or in the pages of Neuro-Psychoanalysis (Solms 1999).

paradigms, as suggested by Giampieri-Deutsch (personal communication). In the future, progress will not be accomplished without larger groups of specialists working together.

The Pragmatism I refer to has been recently explicated by Goldberg (2002)[28]. Practical approaches appeal whenever the subject matter of research interest reaches a critical level of complexity. This is surely the case for our current interest in studying mind or brain both in depth and in tandem. On the other hand, as suggested by Gedo (personal communication), there is also a place for creative thinking in other modes where, for example, discovering the deepest truths is the primary consideration. Combining the search for deep truths with practical real time experimentation and verification may be the best of both worlds.

When we recognize the limitations of our current knowledge, and the obvious benefits of collaborating across disciplinary boundaries, both personally and professionally, we gain in strength, and slowly expand our horizons about such phenomena as memory, emotions, and learning. Therefore, even though we cannot be entirely sure in what way emotional signals emanate from the mind/brain and serve as a critical trigger for the creation of new hope and useful knowledge, we cannot afford to give up our efforts towards eventually discovering more of their technical details. These details obviously lie in our describing more clearly as many as possible of our current unclarities (die "Unklarheiten"), which I hope you will agree I have succeeded in presenting to you. Only then does Night become brighter, as when in analysis we do our best to connect with a fellow sufferer in a compassionate way, and hope this helps him or her also better connect with themselves.

---

[28] Incidentally, since my reference to Pragmatism may undoubtedly confuse some readers, let me briefly expand on what I intend here. I am especially thinking of Charles Sanders Pierce's original thinking about Pragmatism, that is, that the basis of truth determination is via experimental pragmatic results, and I am much less inclined towards William James' and others later modification of Pierce's intentions in the direction of pragmatic subjectivity, or mere efficacy (see Rescher 1995). I also intend the idea that scientists make critical practical decisions in the discovery phase of any investigation, which involves deciding, for example, how to divide complex problems into subproblems that will yield more easily to experimental study.

REFERENCES

Abel, T., Martin, K.C., Bartsch, D. & Kandel, E.R. (1998). Memory suppresser genes: Inhibitory constraints on the storage of long-term memory. *Science* 279: 333–341.

Ahlquist, P. (2002). RNA-dependent RNA polymerase, viruses, and RNA silencing. *Science* 296: 1270–1273.

Albright, T.D., Jessell, T.M., Kandel, E.R. & Posner, M.I. (2000). Neural science: A century of progress and the mysteries that remain. *Cell* 10: S1–S55.

Bartsch, D., Ghirardi, M., Skehel, P.A., Karl, K.A., Herder, S.P., Chen, M., Bailey, C.H. & Kandel, E.R. (1995). Aplysia CREB2 represses long-term facilitation: relief of repression converts transient facilitation into long-term functional and structural change. *Cell* 83: 979–992.

Bartsch, D., Casadio, A., Karl, K.A., Serodio, P. & Kandel, E.R. (1998). CREB1 encodes a nuclear activator, a repressor, and a cytoplasmic modulator that form a regulatory unit critical for long-term facilitation. *Cell* 95: 211–223.

Beattie, E.C., Stellwagen, D., Morishita, W., Breshahan, J.C., Ha, B.K., von Zastrow, M., Beattie, M.S. & Malenka, R.C. (2002). Control of synaptic strength by glial TNFalpha. *Science* 295: 2282–2285.

Bhalla U.S. & Iyengar R. (1999). Emergent properties of networks of biological signaling pathways. *Science* 283: 381–7.

Bliss, T. (1998). The physiological basis of memory. In *From Brains to Consciousness?: Essays on the New Sciences of the Mind*, ed. S. Rose. Princeton: Princeton University Press, pp. 73–93.

Bloom, F.E. (1985). CNS plasticity: A survey of opportunities, In *Central Nervous System Plasticity and Repair*, ed. A. Bignami, A. Adoloye, F.E. Bloom & C.L. Bolis. New York: Raven Press, pp. 3–51.

Bush, G., Phan, L. & Posner, M.I. (2000). Cognitive and emotional influences in anterior cingulate cortex. *Trends in Cognitive Sciences* 4: 215–222.

Castillo, P.E., Schoch, S., Schmitz, F., Südhof, T.C. & Malenka, R.C. (2002). RIM1alpha is required for presynaptic long-term potentiation. *Nature* 415: 327–330.

Contractor, A., Rogers, C., Maron, C., Henkemeyer, M., Swanson, G.T. & Heinemann, S.F. (2002). Trans-synaptic Eph receptor-ephrin signaling in hippocampal mossy fiber LTP. *Science* 296: 1864–1869.

Csikszentmihalyi, M. (1975). *Beyond Boredom and Anxiety: The Experience of Play in Work and Games*. San Francisco: Jossey-Bank.

Dobrunz, L.E. & Garner, C.C. (2002). Priming plasticity. *Nature* 415: 277–278.

Eiseley, L. (1971). *The Night Country*. New York: Charles Scribner's Sons/Macmillan Publ. Co.

Fonagy, P. (1998). Moments of change in psychoanalytic theory: Discussions of a new theory of psychic change. *Infant Mental Health Journal* 19: 346–353.

Ghosh, A. (2002). Learning more about NMDA receptor regulation. *Science* 295: 449–451.

Goldberg, A. (2002). American pragmatism and American psychoanalysis. *Psychoanalytic Quarterly* 71: 235–250.

Gould, E., Reeves, A.J., Graziano, M.S. & Gross, C.G. (1999). Neurogenesis in the neocortex of adult primates. *Science* 286: 548–552.

Hellmuth, L. (2001). Glia tell neurons to build synapses. *Science* 291: 569–570.

Hodgkin, A.L., Huxley, A.F. & Katz, B. (1952). Measurement of current-voltage relations in the membrane of the giant axon in Loligo. *Journal of Physiology* 116: 424–448.

Ibelgauft, H. (1999). Cytokines Online Pathfinder Encyclopedia. http://www.copewithcytokines.de.

Ito, M. (2001). Cerebellar long-term depression: characterization, signal transduction, and functional roles. *Physiological Reviews* 81: 1143–1195.

Johnston, D. (1997). A missing link? LTP and learning. *Science* 278: 401–402.

Kaplan-Solms, K. & Solms, M. (2000). *Clinical Studies in Neuro-Psychoanalysis: An Introduction to a Depth Neuropsychology*. New York: Karnac Books.

Khan, A.A., Bose, C., Lung, S.Y., Soloski, M.J. & Rupp, F. (2001). Physiological regulation of the immunological synapse by agrin. *Science* 292: 1681–1686.

Kohara, K. & Tsumoto, T. (2001). Activity-dependent transfer of brain-derived neurotrophic factor to postsynaptic neurons. *Science* 291: 2419–2423.

Kohara, K. & Tsumoto, T. (2002). Annual Report of Osaka University: Subreport of Research Activities. http://www.osakau-u.ac.jp/eng/research/graphics/.

Levin, F.M. (1962). The effect of epinephrine and 3'5'AMP on the perfused rabbit liver. *The Physiologist* 5: 175.

Levin, F.M. (1991). *Mapping the Mind: The Intersection of Psychoanalysis and Neuroscience*. Hillsdale: The Analytic Press.

Levin, F.M. (1995). Psychoanalysis and knowledge. Part 1: The problem of representation and alternative approaches to learning. *The Annual of Psychoanalysis* 23: 95–114.

Levin, F.M. (1997). Integrating some mind and brain views of transference: the phenomena. *Journal of the American Psychoanalytic Association* 45: 1121–1151.

Levin, F.M. (2002). The neuroimmune network and its relevance to psychoanalysis. *Psychoanalytic Quarterly* 71 (3).

Levin, F.M. (2003). *Psyche and Brain: The Biology of Talking Cures*. Madison: International Universities Press.

Linden, D.J. & Routtenberg, A. (1989). The role of protein kinase C in long-term potentiation: a testable model. *Brain Research Reviews* 14: 279–296.

Lisman, J.E. & Fallon, J.R. (1999). What maintains memories? *Science* 283: 339–340.

Maier, S.F. & Watkins, L.R. (1998). Cytokines for psychologists: Implications of bidirectional immune-to-brain communication for understanding behavior, mood, and cognition, *Psychological Review* 105: 83–107.

Malenka, R.C. & Nicoll, R.A. (1999). Long-term potentiation – A decade of progress? *Science* 285: 1870–1874.

Mauch, D.H., Nägler, K., Schumacher, S., Göritz, C., Müller, E.-C., Otto, A. & Pfrieger, F.W. (2001). CNS synaptogenesis promoted by glia-derived cholesterol. *Science* 294: 1354–1357.

Nakagawa, T. & Sheng, M. (2000). A stargazer foretells the way to the synapse. *Science* 290: 2270–2271.

Panksepp, J. (1998). *Affective Neuroscience: The Foundation of Human and Animal Emotions*. Oxford, New York: Oxford University Press.

Panksepp, J. (2000). The neurodynamics of emotions: An evolutionary-neurodevelopmental view, In *Emotion, Self-Organization, and Development*, ed. M.D. Lewis & I. Granic. New York: Cambridge University Press, pp. 236–264.

Panksepp, J. (2001). The neuro-evolutionary cusp between emotions and cognitions. *Evolution and Cognition* 7: 141–163.

Pfrieger F.W. & Barres, B.A. (1997) Synaptic efficacy enhanced by glial cells. *Science* 277: 1684–1687.

Rescher, N. (1995). Pragmatism. In *The Oxford Companion to Philosophy*, ed. T. Honderich. Oxford, New York: Oxford University Press, pp. 710–713.

Schacter, D.L. & Buckner, R.L. (1998). Priming and the brain. *Neuron* 20: 1–20.

Schoch, S., Castillo, P.E., Jo, T., Mukherjee, K., Geppert, M., Wang, Y., Schmitz, F., Malenka, R.C. & Südhof, T.C. (2002). RIM 1 alpha forms a protein scaffold for regulating neurotransmitter release at the active zone. *Nature* 415: 321–326.

Shi, S.-H. (2001). AMPA receptor dynamics and synaptic plasticity. *Science* 294: 1851–1852.

Shimizu, E., Tang, Y.-P., Rampon, C. & Tsien, J.Z. (2000). NMDA receptor-dependent synaptic reinforcement as a crucial process for memory consolidation. *Science* 290: 1170–1173.

Solms, M. (1999). The new neuropsychology of sleep. *Neuro-Psychoanalysis* 1: 183–195.

Squire, L. (1998). Memory and brain systems. In *From Brains to Consciousness?: Essays on the New Sciences of the Mind*, ed. S. Rose. Princeton: Princeton University Press, pp. 53–72.

Stahl, S.M. (2000). Molecular neurobiology for practicing psychiatrists. Part 5: How a leucine zipper can turn on genes – immedite-early genes activate late-gene expression in the brain. *Journal of Clinical Psychiatry* 61: 7–8.

Takasu, M.A., Dalva, M.B., Zigmond, R.E. & Greenberg, M.E. (2002). Modulation of NMDA receptor-dependent calcium influx and gene expression through EphB receptors. *Science* 295: 491–495.

Wilson, R.I. & Nicoll, R.A. (2002). Endocannabinoid signaling in the brain. *Science* 296: 678–682.

Zacharias, D.A., Violin, J.D., Newton, A.C. & Tsien, R.Y. (2002). Partitioning of lipid-modified monomeric GFPs into membrane microdomains of live cells. *Science* 296: 913–916.

FIGURE CAPTIONS

Figure 1. A simplified scheme of the hypothalamic-pituitary axis (HPA). IS and CNS are captured in a snapshot that shows as well a bit of the complexity of the NIN (Ibelgauft 1999).

Figure 2. Signal transduction underlying LTD (Ito 2001, p. 1151).

C. Empirical Research Trends in Psychoanalysis

X. OVERCOMING ADVERSITY IN ADOLESCENCE: NARRATIVES OF RESILIENCE

STUART T. HAUSER

> "Nothing is so fascinating or complicated as a trajectory of a human life. We emerge partly programmed at birth, and we change with our experiences thereafter. Some of us finally blow apart in adulthood like long-fuse time bombs0, while others grow to shine brightly like comets. Most of us have less spectacular careers, which are still hard to explain in hindsight, even to ourselves, and impossible to foresee in detail."
>
> Jared Diamond (1995, p.16)

1.

Jared Diamond (1995) offers this reflection in his review of E.O. Wilson's autobiography (Wilson 1994), a *narrative* highlighting the daunting challenges involved in understanding the course of individual lives. A distinguished biologist, Wilson describes his early origins as ones filled with major risks: the only child of parents divorced when he was seven, leaving him in the care of others; his father's debilitating alcoholism and eventual suicide; impaired hearing as a child; and in the year of his parent's divorce becoming virtually blind in one eye. Wilson's striking turn from early misfortune to stunning competence and success is a compelling feature of human life drawing the attention, curiosity, and extended energies of many of us to longitudinal studies. In our research program, we are using longitudinal narratives and new narrative analyses to discover basic mechanisms and pathways underlying the kind of surprising life course changes that Wilson chronicles, exemplifying resilient development.

Resilience refers to unexpected adaptation in the face of serious adversity. How such development occurs remains an unsolved mystery (Table 1).

Table 1. *A Central Question Unterlying Studies of Resilience*
What is it about some children and adolescents, their families, or their larger environments that allows them to maintain or subsequently achieve a positive developmental trajectory when many of their peers in similar circumstances are not able to do so? (Adapted from Roosa 2000)

Resilience has picked up many meanings in our popular culture – competence, resilience perfume, resilient marijuana plants, resilient politicians, and the supposed-invulnerable child. Masten and colleagues (1990) point out that resilience can refer to three major classes of phenomena in the psychological literature. Different research approaches are associated with investigations of each set of phenomena:

1) Individuals in *high-risk groups* who have better-than-expected outcomes ("those who overcome the odds" against good development). Theoretical accounts and vignettes about successful people from highly disadvantaged backgrounds are of much interest to many psychiatric and social science observers (e.g., Beardslee & Podorefsky 1988; Harrington & Boardman 1997; Wang & Gordon 1994). Stories of unexpected life paths are often consistent with findings generated by studies of specific outcomes in groups of high-risk children. By pursuing predictors of good outcomes, these correlational and multiple regression analyses shed light on protective factors that might lead to such outcomes (Figure 1).

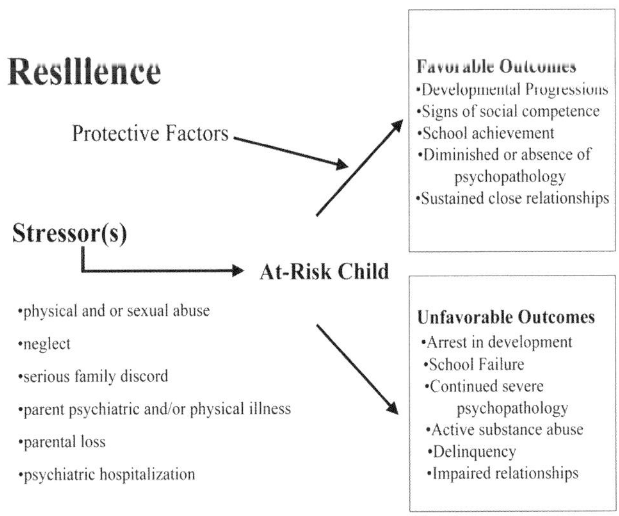

Figure 1

2) *Good individual adaptation despite* adverse events, experiences of misfortune. Sometimes these studies of specific individuals focus on a presumed serious challenge, such as parental divorce. Other investigators target the composite of heterogeneous adverse experiences during a

specific time period in development, like adolescence. Yet a third variation is where one traces the individual's early responses and subsequent development following a particular misfortune, like the early loss of a parent, serious parental illness, sustained poverty. This second perspective leads to an individual life trajectory approach.

3) Individual differences in *recovering from trauma*. Traumatic experiences represent adversities of great severity, with acute onset or chronic repetition (as in child abuse), going well beyond the challenges normally faced in development. Traumatic events or experiences may be natural ones, as in floods or earthquakes; or created by human design, as in war or torture. By their very nature, traumatic experiences are expected to reduce the quality of functioning. In the case of these extreme or life threatening stressors, resilience refers to patterns of recovery (Figure 2).

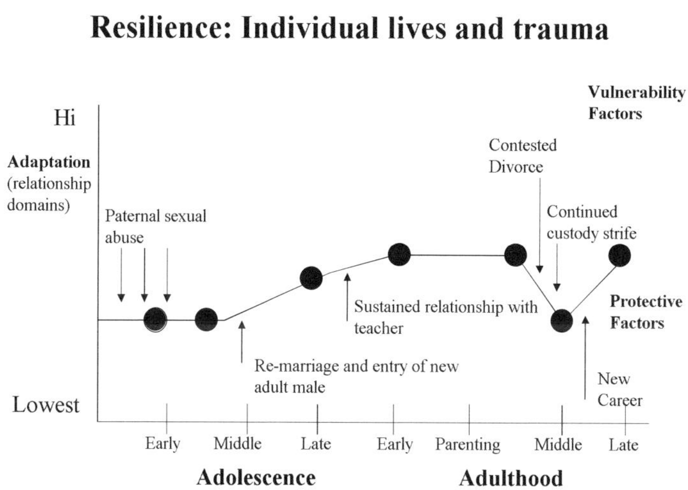

Figure 2

Besides illustrating the third view of resilience, this composite trajectory conveys the idea that an individual's resilience changes over the life course. In our work we use *person-based approaches*, what Luthar and Cushing (1999) call "individual-based measurement" (also, cf. Gjerde, Chang & Kremen 1998). Person-based approaches contrast with variable-based strategies. In variable-based analyses, one follows an (often large) sample of participants presumed exposed to adversity, but missing is specific information on the degree of exposure and experience of each individual to the hypothesized risk factors in such an overall high-risk sample. Consequently, we do not know which individuals within a particular sample actually meet the definition of resilience, specifying *both* high risk and high competence (Table 2; Luthar, Cicchetti & Becker 2000). For all three classes of resilience phenomena, authors generally define resilience in terms of successful adaptation despite challenging or threatening circumstances. This adaptation usually includes internal states of well-being, effective functioning in the environment, or both (Masten et al. 1990, p. 426). The definition of resilient development underlying our studies described below is based on the contemporary one delineated by Luthar and colleagues (2000), refers to internal states as well as observed behaviors.

Table 2. *Resilience* (Luthar, Cicchetti & Becker 2000)

- Resilience refers to a dynamic process encompassing positive adaptation within the context of significant adversity.
- Implicit within this definition are *two critical conditions*
- *Exposure* to significant threat or severe adversity; and
- 2. The achievement of *positive adaptation despite major assaults* on developmental processes.

Protective factors are key constructs within theoretical and empirical studies of resilience. These dimensions moderate the effects of individual vulnerabilities or environmental hazards, so that a given developmental trajectory reflects more adaptation in a particular domain than would be the case if protective processes were not operating. Longitudinal and cross-sectional studies have illuminated protective factors from several domains. Tables 3 &4, synthesized from Masten (1996), Vaillant (1993), Rutter (1987), and Werner (1996; Werner & Smith 1992), summarize processes conceptualized as fostering resilience.

Table 3. *Protective Factors (1)*

Individual	*Relational (including family)*
• Attributional Style • Temperament • *Good intellectual skills • *Appeal to others, especially adults • *Ability to internalize social supports • *Arena of talent or performance valued by self or others • Hopefulness, Faith • *Self-efficacy, Self-esteem	• Effective parenting • *Supportive home environment • *Connections to other • competent adults • *Positive relationships with extended (* = narrative data relevant)

Table 4. *Protective Factors (2)*

Community	*General*
• *Religious Faith or affiliation • *Good schools and other community assets (e.g., clubs, teams)Socioeconomic advantage	(connecting to all domains) • *Good fortune • *Timing (* = narrative data relevant)

Our overarching goal is to more fully understand how resilient development unfolds. Through a person-based approach, guided by a *follow-back* design (Figure 3), we analyze previously recorded adolescent and adult interviews of now competent young adults who experienced significant adversity during their adolescent years. As teenagers, they encountered three serious misfortunes. These young adults were psychiatrically hospitalized from two months to two years during their middle adolescence. Among the consequences of this hospitalization were: First, an abrupt separation from their families, friends and communities. And second, experiencing a serious psychiatric disorder leading to hospitalization, regardless of how time-limited, can markedly change the experience of self, often leading to lowered self regard and lowered perceived personal competence (Cohler et al. 1995). The label of psychiatric patient is made even more indelible by psychiatric hospitalization. The third serious misfortune was trauma. Several of these young adults, now functioning in highly competent ways, reported serious child and adolescent physical abuse at the hands of immediate family members or other close relatives.

As we now study these resilient young adults, we consider individual and interpersonal characteristics that may have acted as protective factors in their lives. Using their *narratives*, drawn from annual clinical interviews in adolescence, we are locating formal and thematic components found in the adolescent discourse of young adults who followed resilient trajectories. While our findings can be most precisely understood as representing individual protective processes, the reciprocal connection between individual and relational protective factors suggests that these narratives will also reveal how the adolescents and young adults recruited, sustained, and experienced *relationships*. In short, our quest to account for young resilient outcomes is drawing upon new analyses of individual narratives expressed over a 20-year period.

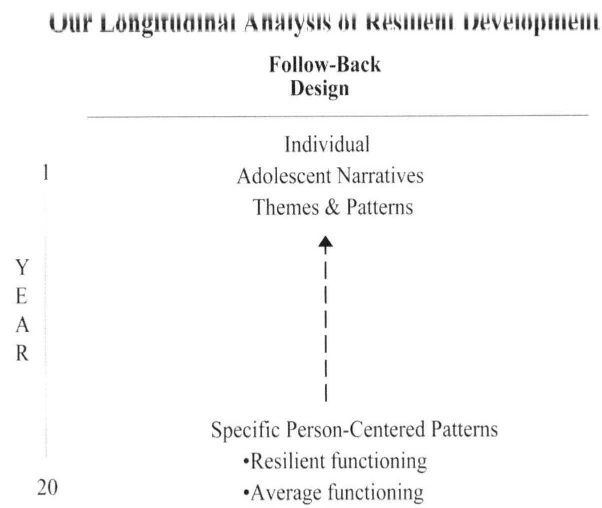

Figure 3

2.

I begin with Eve[1]:

I know I could try. But I know if I try I know how to go back... In between you can go up or you can go down. And if you go up and somebody pushes you down, you're gonna be bummed. And if you go down and nobody helps you back up, you're gonna be bummed. So you should just sit in the middle for now, just see how things work out. And if I feel like I'm gonna talk, I'm gonna. If I don't feel like it, then I'm gonna sink down. Right now I'm sinking. But I think I might be able to work things out.

As a 14-year old, Eve is thinking aloud about changing, a year after slashing her forehead and heavily abusing drugs. Sixteen years later, now a mother of three very young children, she tells about her struggles as a parent:

I'm a good parent... When the kids were younger, I always used to like arts and crafts with them.... And then I ... went back to work. Before I would do a lot of stuff with them, and now it's more like "just find something for yourself to do". . . . . Once in a while I'll get feeling real guilty, and now I'm gonna do something with my kids and I don't care if the house is falling apart. Then so be it. I'm going to do something and we'll go out... We'll sit there and I'll make picnic lunches and we'll get a blanket. And we'll lay on the grass before I've got to go to work.

A decade after first meeting Eve, we discovered that she and eight other young men and women, seriously disturbed during their teenage years, were now leading competent and productive lives. As adolescents they experienced a remarkable derailment. Usually by surprise, they were suddenly admitted to a psychiatric hospital. Separated from their families, in new bewildering surroundings they had to respond to unfamiliar adults whom they never asked to be their caretakers, and attend a

[1] All names of subjects in this paper are not their actual names. In addition, to protect confidentiality, other possible identifying information such as hospital names or locations, are also changed.

special school they found repugnant. Picture these young adolescents: already seriously troubled, they were now thrust into a new disturbing neighborhood, a large hospital community, and a new school; they were forced to contend with many frightening peers, classmates, and adults. How did these teenage boys and girls make sense of this unexpected long stay at High Meadows, a hospital not of their choosing? And now, almost two decades later, how do these young men and women understand their time in the hospital and what happened since their departure?

3.

Our longitudinal resilience project takes up these and related questions (Figure 4). In 1976, we began meeting with 146 middle adolescents *and* their families (Hauser 1991b).

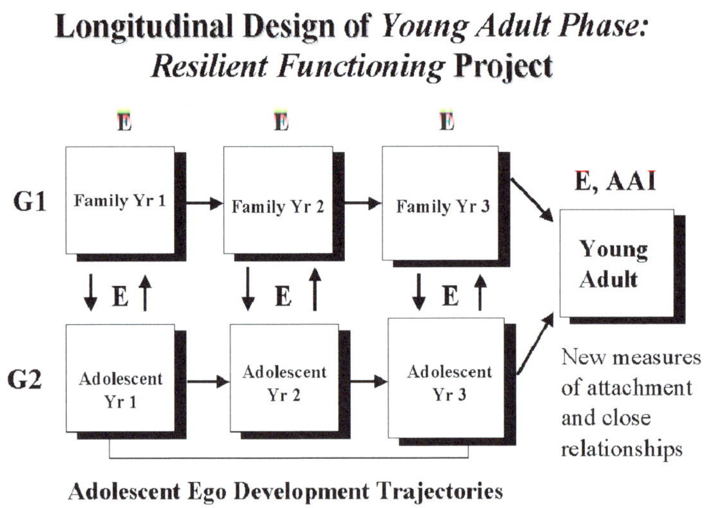

Figure 4

Our sample was drawn from two groups: non-psychotic patients from a private medical school teaching hospital, and volunteers from the freshman class of a local high school (Table 5). The patients included three major diagnostic groups: disruptive behavior disorders, mood disorders, and personality disorders. We excluded psychotic patients, those with evidence of mental retardation, or with medical conditions associated with psychiatric sequelae. The patient and high school samples were comparable in age, race, and family type (one-parent and two-parent); they were predominantly white middle and upper-middle class.

Table 5. *Original Adolescent Sample*

	Male	Female	Total
Psychiatric	39	31	70
High School	34	42	76

Over successive adolescent years, we continued to meet with 85 percent of these adolescents and their families. The adolescent participants responded to specific developmental and personality measures, and participated in annual clinical research interviews usually with the same clinically trained interviewer (Table 6). During these one to two hour clinical research interviews participants were encouraged to associatively respond to inquiries about family and individual history, peer and family experience, school life, emotion regulation, and visions of the future.

Table 6. *Adolescense: Annual Assesments*

1. Ego Development (SCT)
2. Adaptive strengths and defenses (Clinical research interview and coping scales)
3. Emotion Expression (Clinical research interview and AECS)
4. Narrative accounts of self-themes, relationships, and resolutions (Clinical research interview and Readers' Guide)
5. Self-image integration and continuity (Self-image Q-sort)
6. Self-esteem (Coopersmith self-esteem inventory)

In previous analyses we've used the tapes and transcriptions of these interviews to assess defenses, self-images, and emotion expression (e.g., Hauser, Borman et al. 1991; Hauser, Powers et al. 1983; Safyer & Hauser 1994). Now we're examining our original participants' interviews in a new way, looking closely at the *stories* they told over four adolescent years and one young adult year.

Eleven years ago, we located *all* of the former high and low-risk adolescents, and met with 97.3 percent of those young adults still alive (Table 7).

Table 7. *Young Adult Sample*

	Male	Female	Total
Former Psychiatric	37	29	66
Former High School	34	42	76

In this new phase, using a range of measures conceptually relevant to young adult tasks, we assessed the participants' ego development, close relationships, attachment representations, and several aspects of social competence (Table 8). Many of our participants were beginning new families, alerting us to the opportunity for extending our longitudinal program to study marriage and parenting in a third generation.

Table 8. *Dimensions Used to Identify Resilient Young Adult Functioning*

1. Ego Development
2. Quality of Peer Interactions (perceived by friend)
3. Close Peer Relationships
4. Adult Attachment Representation Coherence
5. Social Competence: Criminal Behavior, Substance Abuse & Psychiatric Symptoms

After meeting with the young adults, several interviewers were astonished when the person whom they had been speaking with for the past couple of hours suddenly told them (almost always toward the end of the interview) about their disrupted adolescence, about how they had lived many months or several years in a psychiatric hospital. Our interviewers' amazement was especially meaningful, since they were deliberately not given any information about the histories of anyone with whom they were meeting.[2] These surprises led to our new questions and curiosity about resilience.

What was it about these young adults that surprised the interviewers? Interviewing our interviewers and exploring the resilience literature led to our conceptualizing resilient young adult development in terms of *individual outcome profiles*, encompassing ego development, close relationship functioning, and social competence. We translated this concep-

[2] One exception to this practice was when interviewers were scheduled to meet with former patients who were now in prison. The young woman or man who might interview the prisoner were now told of his psychiatric history (all were males), and given the option of declining the assignment for that interview. No interviewers refused to meet with any of these former patients.

tual definition into an empirical one. What's most important is that through this definition we could systematically identify those young adults whom we classify as *resilient.* Having located this special group, we then used a narrative approach to intensively study the adolescent experience of each participant, thereby tackling the question of whether there were clues in their narratives – between ages 14 and 18 – that might have predicted which boys and girls would later show resilient outcomes. In other words, we followed a group of resilient young adults – *back* to their adolescent years (Figure 3).

Our initial challenge was to operationalize what we thought the interviewers were seeing. We first identified former patients with development and relationship scores *above* the 50th percentile for the entire sample, which includes both non-patient and patient participants. We used four measures: ego resiliency as judged by friends (Kobak & Sceery 1988), relationship closeness (Berscheid 1987), attachment representation coherence (Main & Goldwynn 1998), and ego development (Loevinger 1976) (Table 9).

Table 9. *Young Adult Resilient Functioning (1)* Above 50th percentile of entire sample on:
• Ego Resiliency (Q-sort prototype) • Closeness Inventory • Coherence of Attachment Representations • Ego Development item sum score

We then considered social deviance and psychopathology. To be identified as highly competent on these measures, former patients must have scores *below* the 50th percentile for the entire sample. We used three indices here: hard drug use in the past six months (Elliot et al. 1983), criminal behavior in the past six months (Elliot et al. 1983), and global psychiatric symptoms (Derogatis 1983) (Table 10).

Table 10. *Young Adult Resilient Functioning (2)* Below 50th percentile of entire sample on:
• Hopkins Symptom Checklist • Delinquency/Crime Questionnaire • Substance Abuse Questionnaire

Four men and five women, thirteen percent of the 67 former patients, fit this profile definition of young adult resilience.

Having identified the young adult resilient subjects, we could now – through our follow-back design – examine our years of adolescent and adult observations. Available to us were data generated by a panoply of methods, encompassing semi-structured interviews, self-report questionnaires and systematically coded observed family interactions. Instead, we chose a new point of departure: the *complex and resonant personal narratives* embedded within the many hours of each subject's adolescent and young adult interviews. Through new detailed studies we expected to discover previously uncharted individual and relational protective processes. Because we planned to follow, for each subject, up to eight interviews over two developmental eras, we would also be positioned to probe yet another compelling question: Does the nature and effectiveness of protective processes change over the life course for resilient and less resilient subjects?

4.

Gay Becker (1997) describes narratives as the stories people tell about themselves, reflecting people's experience as they see it and as they wish to have others see it. "Through stories people organize, display and work through their experiences." She further argues that "narratives can be a potent force in mediating disruption, whether the disruption is caused by illness or personal misfortune" (Becker 1997, p. 25). Through narrative analyses we can begin to grasp how individuals create and maintain coherence over time. Rita Charon extends these ideas in her recent writing about narrative medicine. In her words,

"As patient meets physician, a conversation ensues. A story – a state of affairs or a set of events is recounted – by the patient in his or her acts of narrating, resulting in a complicated narrative of illness told in words, gestures, physical findings, and silences and burdened not only with the objective information about the illness but also with the fears, hopes and implications associated with it. As in psychoanalysis, in all of medical practice the narrating of the patient's story is a therapeutically central act, because to find words to contain the disorder and its attendant worries gives shape to and control over the chaos of illness." (Charon 2001, p. 1898)

Narrative approaches have a long history in the social sciences and psychoanalysis. Within psychoanalysis, Schaefer (1992) and Spence

(1982) use narrative frameworks in their considerations of psychoanalytic dialog, constructions and epistemology. Besides Becker and Charon, influencing the narrative methods we're developing are contributions by Bruner (1990, 1994), Mishler (1995), Gilligan (Brown & Gilligan 1992), Lewis (1997), Main (Main & Goldwyn 1998), Reissman (1993), and Cohler (e.g., Cohler, Stott & Musick 1995).

Based on our reading of the resilience and adult attachment literatures, and many reviews of the adolescent interviews, we created a Guide for analyzing narratives within the open-ended interviews[3]. Our guide alerts the narrative analyst to content areas and structural features likely relevant to resilient outcomes. We begin by demarcating *two stories*, the patient's path to the hospital and his or her life within the hospital. These stories are extracted from all of the first year interviews (Table 11). Important formal aspects of these and subsequent key stories are coherence and passivity or vagueness in discourse (Main & Goldwyn 1998). Thematic components of interest include representations of self and of interpersonal relationships. With respect to *self* (Table 11), we consider ways of disclosure and privacy, self-reflection, agency, self-efficacy, self-esteem, aspirations, helplessness, and long-term visions. Representations of *relationships* (Table 11) include the extent to which the person speaks about them as interconnected; how he or she seeks, recruits, and maintains ties with others.

Noteworthy is the fact that we do *not* intend our guide to be a constrained coding system, a rigid template forcing categories onto these rich texts. Rather, our guide is designed to sensitize the reader to important clusters of themes, while encouraging openness to seeing novel patterns within the interview discourse. After the hospital year (Table 11), we continue to analyze subsequent year's interviews, using the same guide. Now the questions are applied to new stories about *current life* and *self-described changes*. But when a participant brings up his path to High Meadows and to his life there, we pay close attention, since we expect that this revisiting may shed new light on the changing ways resilient adolescents made sense of an extraordinary time in their lives.

---

[3] The ideas for this guide were strongly influenced by the work of Brown and Gilligan (1992) and Main and Goldwyn (1998).

Table 11. *Guide for Narrative Analysis of the Adolescent and Young Adult Interview*

1. Identify two sets of stories in year one: a. Path to the hospital b. Life within the hospital
2. Within these stories identify all: *a. Representations of self* i. coherence ii. disclosure and privacy iii. self-regulation iv. mastery v. long-term visions vi. other aspects *b. Representations of relationships* i. degree of interconnectedness, coherence ii. sought and elicited iii. sustained and/or changed iv. as significant resources v. other aspects
3. Identify new stories in each successive year about: *a. current life* *b. self described change*
4. Within these stories, and in new stories about "paths" and "hospital life" identify all: *a. Representations of self* *b. Representations of relationships*

Since I believe that it is through following individual lives over time that we can gain the strongest insights from our narrative analyses, let's return to Eve, tracing her narratives of parenting first as an adolescent (having parents) and then as a young adult (being a parent).

4. 1. Eve

Restless and desperate at school and at home, Eve was admitted to High Meadows Hospital at age 13, acutely confused, with disorganized behavior, and self-mutilation. For years, Eve expressed mounting disappointment about her parents. Although she first described her father as supportive, in all later years she portrayed him as severe, dogged, angry, and disrespectful. She felt deeply troubled by him, in two ways.

My father gave up trying... He told me no matter what, he would never give up.... He let me down and he let himself down . . . I knew all the time my father was going to give up.... And it wasn't a shock to me.... It's like you are waiting for someone to die. When they die, you know.

The second way was by consistently telling her how bad her thinking and friends were, stubbornly insisting that she's an "asshole". ".... [I'm] getting to the point I'm not gonna care. I am gonna end up being a jerk – I'll probably go away."

Yet sometimes, while anguished over her father's verbal abuse and her mother's passive agreement with the abuse, Eve at the same time reflects about them:

…they always blamed their problems on us. We could be a little bit of the cause... But they didn't have to do half the shit they did. The choice was theirs.

Similarly, Eve recognizes that it's her *own* choice to "do anything wrong"...

Because I know what's right and wrong... I'll just make decisions for myself... and if it turns out that I'm an asshole by making a decision instead of him, then that's tough luck... I found out the only person who could fix things is me... If I mess up, at least I will know it's my fault.

Within Eve's reflections about herself is an important quality – recognizing her responsibility for her thoughts and actions, what Bakan (1966) terms human "agency".

In late adolescence, Eve describes more peaceful times with her parents, now liking her mother "better", and not talking to her father because he makes her too uncomfortable. In great detail, she characterizes his fixed ideas; in any discussion with him he "always comes back with his opinion and his opinion would always have to be right...whatever he says goes." What's now fostering greater harmony within the family is Eve's not confronting him by yelling at him. "I'll say [to him] this is what I want to do." Seeing she's like her father in being "dead set against things," is a new insight she has in her last adolescent interview.

Seven years later, when we meet Eve as a young adult, she portrays more complex pictures of both parents *and* continues to speak about themes she expressed in her adolescent narratives. Her mother, as Eve looks back on child and adolescent years, was giving, caring, and understanding. These are clearly more positive attributes than ever appeared in Eve's adolescent narratives. Yet Eve recalls how confused her mother was, "because she didn't know what to do," and how mad she was, because "she was stuck in the house all day with her three children, unable to go out and find a job or return to school."

Eve's young adult narrative of her father first appears more generous – fun loving, caring, kind. In a few moments, her more ominous image appears:

He had a mental way of working on you, [making you] feel....you were worthless...he'd put you down about everything you did...If you cared about your friends, he'd say, "your friends are a piece of garbage," and to us they were important... No physical abuse...just threats...

Now a parent of three children under five years old, Eve describes her almost automatic reaction of quickly closing down and not listening, the moment she senses her father is about to criticize her, to get in one of his states where, "... he's being so ignorant about being not able to budge either way in understanding where you're coming from."

As she looks back, Eve's strongest regrets about her father remain: That he couldn't see where she was "coming from", and not intimidate her. While some of her constructions of him have become more nuanced and positive, her powerful awareness of his stubbornness and verbal abuse persist, presenting constant dangers to her growing sense of confidence and well-being.

Unhappily, Eve tells us about her second child. She sees her four-year old son, Bobby, as "more like his dad, because he's real slow." And not only is he slow, "it's like he's not trying... He gives up more... He'll go back one day and finally get it." Added to his slowness and lack of persistence is his stubbornness, short attention span and provocative temper tantrums. Echoing her experience of pervasive criticism by her father are Eve's strong and many-sided disappointments with Bobby.

Eve refuses to tolerate Bobby's weaknesses. She believes her attitude will help him know that other people are not going to tolerate his limitations either. Her father and husband converge, as she describes her husband as looking at things "like either it's right or wrong and that's not it." In contrast, Eve is proud of her propensity to step back, and think about there being

...different ways, different perspectives and depending on...the way you are in life,...things change so you can't always be contradicting things or criticizing things. You have to be more of an open-minded type. I got that from my teenage years.

Finally, Eve believes that one of her most serious limitations is her reluctance to strongly discipline Bobby. Her parents were also reluctant, and her father still regrets not beating her. She has Bobby go to the corner, take time out, hold his breath. Similar to her parents, who "didn't hit us much," Eve does not want to punish her children in "heavy duty" ways.

Along the lines of content (discipline, specific child criticism of stubbornness, and lack of persistence) and style (being reflective, taking responsibility), we see Eve expressing – over a decade later – varied continuities from her adolescent narratives. Yet we also find signs of change, in terms of her more multifaceted and forgiving descriptions of her parents.

Sharply differing from Eve, Pam and Jenny – average outcome patients – speak in limited ways about their new parenting experiences. Only occasionally do they speak of their own parents. Both young women are having great difficulty as parents – Jenny has lost custody of her only daughter; and Pam is tired of her several marriages and of having children each year. These average outcome former patients tell us of unstable marriages, together with barely decipherable connections with adolescent parenting images. Their stories are stark, portraying great loneliness, few signs of agency or hopes for change, and few relationships.

4. 2. Pete

A second resilient young adult arrived at High Meadows after 12 public and private school expulsions, where his explosive violence made him almost instantly unwelcome in each new setting. Yet with unusual ease for a fourteen-year-old, Pete tells new adults at High Meadows about a family life replete with threats, tempers, and sudden physical brawls. His abusive relationship with his father was a model, from early times, for volatile destructive episodes with others. Pete witnessed seemingly uncontrolled abuse between his father and his siblings, recalling scant protection from his mother. And when his father was drunk, these experiences and the aftermath of regrets were even more dreadful. In the midst of these most turbulent and troubled adolescent years, we hear Pete's lively inner conversations about troubled relationships and desires. He needs friends, desperately at times. And each friend frightens him. Guided by his formidable talent in sizing up himself and others, Pete recruits various adults – whom he calls his 'mentors' – to be his companions. This recruiting talent is one of his greatest assets.

Just before coming to High Meadows, Pete makes his fury towards his father more visible and frightening. Stealing his grandfather's gun, he sends a message to his father. "I didn't tell him directly. But I'm sure he had heard about it. Cause I told enough of his friends and enough of his relatives that I'd kill him if he touched me again." Now a young teenager, Pete has discovered a way to control his father's blazing rage: directly and indirectly threatening his life. Pete now recruits a new mentor, Bob Jenkins, a school social worker, to whom he reveals both his hatred of his father *and* how terrified he is about carrying out his deadly threat.

One time it got so bad and I couldn't hack it any more. And I just went to [this friend] who worked at the school. I said, "You gotta do something...put me anywhere, but away from him. [his father]"... I had a lot of [other worries] too...like my walking a younger kid into the park at knife-point. That's why I went to the emergency care shelter, cause everything was driving me crazy, and I would have killed myself.

During his two years at High Meadows Peter makes much progress in controlling his fury. And then he faces many setbacks: close relationships ending; disappointment and discouragement at an acting school where he imagined becoming a great star; and the threat of alienation from his mother, the most powerful and revered adult in his life. With these accumulating reversals, Pete decides to leave the "state". Under a false identity, he travels across the country, stopping himself from making any strong new relationships. After one year, he returns home, once again valuing relationships.

Let's fast-forward seven years. Pete has now graduated from a major research university, is a writer, and has his first full-time adult job in a computer firm. And he has found a new close friend, Arnie, with whom he lives and travels from late adolescence through his young adult years. Alongside intermittent intense disruptions in their relationship, Pete experiences Arnie as a caring devoted older friend, one more important man is the line of his older male friends, mentors he's been recruiting since before high school:

These were all mentors. These were all the people who had a lot of bearing, a lot of influence on my life. And I know it sounds weird. I don't have a lot of friends my age... I guess I was replacing my father like 10 times over.

In their ten-year stormy friendship, Arnie and Pete have traveled together all over the world. Between trips Arnie rescues Pete from urgent situations, like bailing teenage Pete out of jail after big public fights or drug busts. Over the years they talk about their loneliness, about Pete's becoming "tongue tied" in close relationships with girls. They have intense arguments in airports and on trips, usually triggered by Arnie's jealousy over Pete's new sexual flirtations; or by Pete's idea that Arnie is "playing games" with him, threatening by innuendo or not saying what's really on his mind. Arnie points out to Pete times when he's withdrawing, becoming one more voice pressing Pete to think about his "barriers".

...[Arnie's] one of the people who accuses me of having emotional barriers. And he always wants our friendship to get even deeper, as emotionally. He's not coming on to me... But he always says I keep a barrier, just a certain distance. And it pisses him off.... I was brought up like that. Because my mother was a little like that... My grandfather was a little like this. So he feels it.... He's from a very warm Italian family...people love and hug and cry together. And, Jesus, that just doesn't happen in my family...

Pete's perception of his and Arnie's very different backgrounds can't fully explain the vacillating closeness and distance he experiences in other relationships. He also sees his exquisite fear of being hurt by others leading him to jump back, cut off, and then slowly return to the other person. More clearly becoming aware of his fears over being abused, submitting, and surrendering, is what Pete's referring to when declaring that he's grown emotionally through friendships and "tons and tons and scads of therapy...and I worked very hard to work myself out.… I had more support somewhere." The tension between his appreciating the importance of close friends and his difficulties tolerating close relationships is the central dilemma of Pete's young adult years.

So why, as I describe these continued internal and interpersonal struggles, do I also tell you that Pete represents a resilient former patient father and stepfather. In fact, marking Pete's psychological development is the radical change in his relationship with his stepfather. Once called a sub-human 'computer' with no feelings, a man whose 62-year-old bones Pete wanted to crush, he has now come to be a valued friend. At age 35 Pete told me he was going to Europe to take a bike trip across Belgium with him. I asked him if this was the same stepfather he spoke about in teenage days. And he said, "Oh yes. I've really changed; now we're close friends." Pete illustrates well the idea that resilience is not a monolithic unchanging trait within a person. It is a dynamic patterning of strengths and sensitivities responsive to age and context. This profile of an individual's resilient features changes over the life cycle with respect to its salient dimensions.

5.

Through new narrative analyses of these resilient young adults, we're becoming aware of experiences of self and of relationships that simply weren't tapped through previous empirical procedures. For instance, our more traditional analyses were theory-driven and constrained by rigorous coding conventions and technical language as we were pursuing indices of defenses, adaptive strengths, self-images, expressed effects, and enabling interactions. In working with our previous methods, we often had the impression that we were *not* capturing significant aspects of the subject's experience; especially their vividly described the poignant past and current relationships. And when our coders would bring up these limitations, we'd usually speak – somewhat dismissively – about the compromises forced upon us by systematic empirical research; how we could not possibly study all the unique features embodied in one subject's interview. But our coders were, in hindsight, obviously correct in pointing out a meaningful problem. Their concerns had to do with at least two issues. First, the units we were analyzing were too constricted. Fuller portions of text are required to locate certain nuances of form and meaning. Second, we focused on categories derived from theoretical perspectives we believed relevant to testing our hypotheses. In other words, we were *not* identifying personal meanings, ways a subject was making sense of his or her world. Verghese (2001) speaks of this problem: "Any professional language brings with it the risk that it will also put blinders on us, bring about an atrophy of our imagination, a waning of the ability to understand the suffering of the patient." (Verghese 2001, p. 1015) These concerns led to our constructing a new kind of analytic procedure, our Guide? Across several domains – ego development, attachment coherence, substance abuse, emotion regulation – 25 year old Pete is stunningly different from 16 year old Pete, a bereft and furious boy, rushing to leave the state; a teenage boy with violent maiming fantasies about his for Interview Texts, designed to systematically and reliably extract structural and content dimensions from our semi-structured adolescent and adult interviews.

Based on these explorations, we think there are strong reasons to add narrative analyses to our ongoing life span studies. From our examination of more than 60 adolescent and adult-interviews from the resilient and average outcome young adults – spanning over 20 years – we have a better grasp of resilient participants' constructions of themselves and their relationships during a time of major disruption; and we have seen ways these constructions change over time. In terms of constructions of *self*, we have found five content themes and one structural feature (Table 12):

Table 12. *Resilient Young Adults' Narratives of Self During Adolescence (1)*	
Major features:	
1. Self-reflection (reflective self)	
2. Self-efficacy, or agency	
3. Self-complexity	
4. Persistence and ambition	
5. Vicissitudes in self-esteem ⟶	overall increase in self-esteem
6. Coherence	

1. *Self-reflection*, illustrated by individuals' increasing awareness of their feelings and thoughts, within each adolescent year and in later thinking about their experience and performance as adults and new parents. A few weeks after entering High Meadows Pete was stunned right after grabbing Joey, another student annoying him at school: "... I stopped about *that close* to his face. And I said [to myself], 'What am I doing?...What the hell am I doing?'.. that's not like me any more, to jump up and take that extreme. I usually say something before [I get totally furious] and they'll stop."

2. *Agency.* As adolescents, the resilient patients played an active part in deciding where they would go after leaving High Meadows and how they would take care of themselves when again in trouble. Sandy admitted herself to a community hospital, two years after leaving High Meadows, when she couldn't stop taking 'uppers' and 'downers.' Pete left the

state in order to disrupt the destructive cycles which threatened his maintaining relationships with his mother and friends. As adults, they make conscious choices about parenting and how to put their ideas into practice – recall Eve and her picnics. Related to this theme are the resilient adults' detailed visions of their future, and their conscious plans to make sure they actualize these optimistic dreams.

3. *Self-complexity.* We've seen this feature in Eve's recognition of her many sides. Pete speaks of his complexity as he many times tells us about his multiple worlds – his tough biker pals, his concert crowd. We've found evidence of complex continuities over time, the resilient subjects' recognizing parenting themes from their past; they embrace their discovery of multiple perspectives about themselves, and about their friends.

4. *Persistence and ambition.* Resilient patients describe a refusal to settle for a specific solution offered by a hospital, a therapist, or their family. Several find new schools. Two years after her discharge, Sandy arranges for admission to a community psychiatric facility to help her with a new drug problem, and then discharges herself when she thinks the problem is solved. Others return to complete training many years after interruptions.

5. *Self-esteem.* Ever apparent are the resilient participants' vacillating appraisals of themselves. By no means were these evaluations simply increasingly positive over the years. As we've seen with Pete, they were marked by swings of confidence as well as disappointment, optimism and pessimism about life's chances. Important was their awareness of these self-evaluations, and the overall balance tipping in the direction of kinder self-regard each year.

6. These are *coherent narratives.* From the start, we were struck by the ease of discerning the resilient subjects' stories – first about their paths to High Meadows, then about their experiences there. In subsequent years, they provide coherent accounts of their lives, conveying personal successes and failures, and new views of their hospital experience. Sometimes their "ups and downs" are breathtaking and turbulent. Yet the interviewer, and later the reader, can grasp the alternating disappointments and successes. Changes, and connections to the past were often the very first things a resilient subject talked about each year when he or she met with their interviewer. Audiences hearing their stories are often touched by the alternating disappointments and successes.

Turning to constructions of *relationships*, we see three themes (Table 13): the resilient adolescents do much reflecting about others' motives, feelings, and thoughts. Second, the resilient individuals attribute great importance to close friendships and have a thirst for friendships. They tell us about *recruiting* relationships. They do not imagine simply meeting people by chance. They tell us of how they *find* new friends and how these friends help in their recovery from disrupted teenage years. Recruiting and retaining mentors was a crucial theme through Pete's adolescent and young adult years. The special significance of relationships further unfolds in adulthood, as the resilient participants speak of their new spouses and about parenting with their spouses. Finally, the resilient individuals see many *intersections* among themselves, their relationships, and their actions; speaking, for instance, about how growing good feelings about themselves are leading to more positive ways they're seeking and respecting new friends.

Table 13. *Resilient Young Adults' Narratives of Relationships During Adolescence (2)*
Major features: 1. Reflection about others' motives, feelings, and thoughts ("psychological mindedness") 2. Friends and other close relationships as key resources to maintain and invest in 3. acknowledged intersections among self-representations, relationships, and actions.

In the long run, what can we gain from narrative analyses of sequential adolescent and adult interviews? One key benefit can be greater understanding of how these adolescents' narratives contributed to their subsequent adult development. For example, we could hypothesize that an adolescent's increasing narrative complexity and coherence may herald signs of psychological health in adult years. An additional intriguing possibility is that adolescent patients' conscious recognition of personal change predicts ongoing and subsequent adaptations. As we think about direction of influence, it's important to recognize the strong possibility that causal paths may not necessarily move from narratives to outcomes. These narratives could be markers, reflecting changing life circumstances and optimism, leading to change in narrative understanding; and then to new impacts from the narratives themselves, to actions, as the resilient participant makes plans – like future families and having children.

Through longitudinal interview analyses of these nine resilient patients, alongside parallel analyses of average outcome patients, we can begin to delineate some of these complex causal chains. For instance, our first analyses of average outcome patients suggest that they experienced greater helplessness, rage, and diminished self-esteem as adolescents. Fewer steadily supportive and protective relationships with friends and family were available. Nor did these participants consistently recruit relationships. Their adult marital relationships are volatile. Parenting is neither easy nor cherished. As we identify these contrasting themes, we will take the next step of defining new variables – like reflectiveness, agency, relationship recruiting – for which we will systematically search in the adolescent and adult interviews. We can begin to specify how these narratives may shape important adult outcomes – including close relationships, parenting, and psychopathology.

Understandings generated by this new way of looking at our data should also shed light on how these patients adapted to trying circumstances during a most turbulent adolescence. Tracing the flow of meanings constructed by adolescents and young adults can lead to our locating new individual and relational protective factors. Formal characteristics of their teen-age narratives, like increasing coherence and diminished passive discourse, may be among the special features that distinguish the resilient young adults from the other former patients. Through such developing strengths, these adolescents may have compensated for serious psychopathology and adverse hospital and home circumstances, as well as capitalized on available resources –psychotherapy, special teachers, friends, and schools. Our other instruments may not be attuned to the subtle manifestations of these strengths. We are taking this narrative turn to see how personal meaning and meaning-making can foster unexpected pathways from adolescent misfortune to young adult competence in friendship, marriage, and parenting.

REFERENCES

Bakan, D. (1966). *The Duality of Human Existence.* Boston: Beacon Press.

Beardslee, W.B. & Podorefsky, D. (1988). Resilient adolescents whose parents have serious affective and other psychiatric disorders: Importance of self-understanding and relationships. *American Journal of Psychiatry* 145: 63–69.

Becker, G. (1997). *Disrupted Lives: How People Create Meaning in a Chaotic World.* Berkeley: University of California Press.

Berscheid, E., Snyder, M. & Omoto, A.M. (1987). The relationship closeness inventory: Assessing the closeness of interpersonal relationships. *Journal of Personality and Social Psychology* 57(5): 792–807.

Brown, L. & Gilligan, C. (1992). *Meeting at the Crossroads: Women's Psychology and Girls' Development.* Cambridge, Mass.: Harvard University Press.

Bruner, J. (1990). *Acts of Meaning.* Cambridge, Mass.: Harvard University Press.

Bruner, J. (1994). The narrative construction of reality. In *Psychoanalysis and Development: Representations and Narratives*, ed. M. Ammaniti & D.N. Stern. New York: New York University Press.

Charon, R. (2001). Narrative medicine: A model for empathy, reflection, profession, and trust. *Journal of the American Medical Association* 286: 1897–1902.

Cohler, B.J., Stott, F.M. & Musick, J.S. (1995). Adversity, vulnerability, and resilience: Cultural and developmental perspectives. In *Developmental Psychopathology*, ed. D. Cicchetti & D.J. Cohen. New York: John Wiley and Sons, Inc., pp. 753– 800.

Derogatis, L.R. (1983). *SCL-90 Administration and Scoring Manual.* Towson, MD: Clinical Psychometric Research.

Diamond, J. (1995). Portrait of the biologist as a young man. In *New York Review of Books*. January 12.

Elliot, D.S. & Ageton, S.S. (1980). Reconciling race and class differences in self-reported and official estimates of delinquency. *American Sociology Review* 45: 95–110.

Elliot, D.S. Ageton, S.S., Huinzega, D., Knowles, B.A. & Canter, R.J. (1983). *The Prevalence and Incidence of Delinquent Behavior: 1976–1980.* Boulder, CO: Behavioral Research Institute.

Gjerde, P., Chang, R. & Kremen, A. (1998). Life paths through adolescence: A study of developmental prototypes. Presented at Seventh Biennial Meeting of the Society for Research on Adolescence, San Diego, CA, February 27, 1998.

Harrington, C.C. & Boardman, S.K. (1997). *Paths to Success: Beating the Odds in American Society.* Cambridge, MA: Harvard University Press.

Hazan, C. & Shaver, P. (1987). Romantic love conceptualized as an attachment process. *Journal of Personality and Social Psychology* 52: 511–524.

Hauser, S.T., Jacobson, A., Noam, G. & Powers, S. (1983). Ego development and self-image complexity in early adolescence: Longitudinal studies of diabetic and psychiatric patients. *Archives of General Psychiatry* 40: 325–332.

Hauser, S.T., Borman, E., Bowlds, M.K., Powers, S., Jacobson, A., Noam, G. & Kroebbler, C. (1991). Understanding coping with adolescence: Ego development and coping styles. In *Life Span Developmental Psychology: Perspectives on Stress and Coping*, ed. A.L. Greene, E.M. Cummings & K. Karraker. Hillsdale, NJ: Erlbaum Associates, pp. 177–194.

Hauser, S.T., Powers, S. & Noam, G. (1991). *Adolescents and Their Families: Paths of Ego Development*. New York: Free Press.

Hy, L.X. & Loevinger, L. (1986). *Measuring ego development*. Mahwah, NJ: Erlbaum Associates.

Kobak, R. & Sceery, A. (1988). Attachment in late adolescence: Working models, affect regulation, and representations of self and others. *Child Development* 59: 135–146.

Lewis, M. (1987). *Altering Fate: Why the Past Does Not Predict the Future*. New York: Guilford.

Loevinger, J. (1976). *Ego Development: Conceptions and Theories*. San Francisco: Jossey-Bass.

Luthar, S.S., Cicchetti, D. & Becher, B. (2000). The construct of resilience: A critical evaluation and guidelines for work. *Child Development* 71: 543–562.

Luthar, S.S. & Cushing, G. (1999). Measurement issues in the empirical study of resilience: An overview. In *Resiliency and Development: Positive Life Adaptations*, ed. M. Glantz, Z. Sloboda & L.C. Huffman. New York: Plenum, pp. 129–160.

Main, M. & Goldwyn, R. (1998). Attachment scoring and classification systems (Version 6.1). Unpublished manual. Berkeley, University of California.

Masten, A., Best, K. & Garmezy, N. (1990). Resilience and development: Contributions from the study of children who overcame adversity. *Development and Psychopathology* 2: 425–444.

Masten, A.S. (1996). In *Fostering Resiliency in Kids: Overcoming Adversity*. A congressional breakfast seminar. Consortium of Social Science Associations, Washington, D.C., pp. 19–24.

Mishler, E.G. (1995). Models of narrative analysis: A typology. *Journal of Narrative and Life History* 5(2): 87–123.

Reissman, K. (1993). *Narrative Analysis*. Newbury Park, CA: Sage.

Roosa, M. W. (2000). Some thoughts about resilience versus positive development, main effects versus interactions, and the value of resilience. *Child Development* 71: 567–569.

Russell, D., Peplau & Cutrona, C. (1980). The revised UCLA loneliness scale: Concurrent and discriminant validity evidence. *Journal of Personality and Social Psychology* 39: 472–480.

Rutter, M. (1987). Psychological resilience and protective mechanisms. *American Journal of Orthopsychiatry* 57: 316–331.

Safyer, A. & Hauser, S.T. (1994). A microanalytic method for exploring adolescent emotion expressions. *Journal of Adolescent Research* 4: 487–502.

Sameroff, A.J., Seifer, R. & Zax, M. (1982) Early development of children at risk for emotional disorder. *Monographs of the Society for Research in Child Development* 47(Series No. 199).

Schafer, R. (1992). *Retelling a Life: Narration and Dialogue in Psychoanalysis*. New York: Basic Books.

Schultz, L.H. & Selman, R.L. (1998). Ego development and interpersonal development in young adulthood. In *Personality Development*, ed. P.M. Wesbenberg, A. Blosi & L. Cohn. Mahwah, NJ: Erlbaum, pp. 181–202.

Spence, D. (1982). *Narrative Truth and Historical Truth: Meaning and Interpretation in Psychoanalysis*. New York: Norton.

Vaillant, G. (1993). *The Wisdom of the Ego*. Cambridge (MA): Harvard University Press.

Verghese, A. (2001). The physician as storyteller. *Annals of Internal Medicine* 135: 1012–1017.

Wang, M.C. & Gordon, E.W., ed. (1994). *Educational Resilience in Inner-City America: Challenges and Prospects*. Hillsdale, NJ: Lawrence Erlbaum Associates.

Werner, E.E. (1996). In *Fostering Resiliency in Kids: Overcoming Adversity*. A congressional breakfast seminar. Consortium of Social Science Associations, Washington, D.C., pp. 6–14.

Werner, E.E. & Smith, R.S. (1992). *Overcoming the Odds*. Ithaca (NY): Cornell University Press.

Wilson, E. O. (1994). *Naturalist*. Washington, DC: Island Press/Shearwater Books.

FIGURE CAPTIONS

Figure 1. Resilience.

Figure 2. Resilience. Individual lives and trauma

Figure 3. Longitudinal analysis of resilient development. Follow-back design

Figure 4. Longitudinal design of young adult phase. Resilient functioning project

XI. STRUCTURAL CHANGE AND ITS ASSESSMENT. EXPERIENCES FROM THE STOCKHOLM OUTCOME OF PSYCHOANALYSIS AND PSYCHOTHERAPY PROJECT

ROLF SANDELL

If we were to launch a survey among psychoanalysts of different schools in different societies about the goals of psychoanalysis, my hypothesis is that a large majority, irrespective of school or society, would include some concept of structural change among the most important ones, if not as *the* most important. Time and again when I have presented our findings to psychoanalytic audiences the question has been asked if – or, why – we did not measure "structural change". So it is partly in response to that question from a couple of colleagues on that occasion that I take the opportunity to offer in this chapter a more extensive answer than I was able to do then.

1. HOW HAS STRUCTURAL CHANGE BEEN DEFINED?

I am not sure who first used the concept of *structural change* in psychoanalytic discourse. I have found no entry for it in *The Language of Psycho-Analysis* (Laplanche & Pontalis 1980) and none for any reasonable synonym or alternative rendering. Correspondingly, it seems to be missing in the General Subject Index in the *Standard Edition* of Freud's (Freud 1974) work. One may assume that is was implied for the first time in the famous formula "Wo Es war, soll Ich werden." However, there may have been structural change notions around much earlier, in terms of the topographical model. In view of the fact that Freud, at the time, considered the conscious and the unconscious as systems, the formulation to make unconscious conscious may be taken to imply, too, the idea of systemic or structural change.

Nowadays, the concept of mental structural change has become what Weinshel (1990) has called a psychoanalytical shibboleth, a slogan, a rhetoric to distinguish psychoanalytic change from the – presumably less genuine – forms of change in other kinds of mental treatments. But Weinshel argues (1990, p. 636), "We should at least consider that our long-held claim that 'structural change' is the sacrosanct province of psychoanalysis may be a somewhat solipsistic one." My purpose in this paper is to follow Weinshel's suggestion and analyse the structural change concept with the following kinds of questions in mind:

What are the essentials of the kind of change that may be called structural (in contradistinction from non-structural kinds of change)?

Is the structural kind of change specific to psychoanalysis or may it come about in other ways?

Are all changes during a psychoanalysis structural?

How may structural change be properly assessed?

2. THE NOTION OF STRUCTURAL CHANGE IN PSYCHOANALYSIS

The easy way to define structural change is to claim that it is the kind of change that occurs in a psychoanalysis. This is a primitive definition, of course, more or less a stipulative type of definition, saying merely that whatever change occurs in a psychoanalysis is a structural one. A variation and, in fact, much stronger rendering of the idea is that structural change is the kind of change that may occur *only* in a psychoanalysis. At once, this becomes a bit more complicated, because this is a reality type of definition, saying that there is something, which we shall call structural change, that will happen in psychoanalysis, *exclusively*. By being a reality definition, saying something about structural change, it may be tested against reality. One thing that has been found in reality is that, whether it is specific to psychoanalysis or not, there are not always changes observed in psychoanalysis that analysts would like to call structural change. So, therefore some have chosen to link the concept to another concept, that of psychoanalytic process, and argue that structural change is the kind of change that occurs – may *only* occur, possibly also occur *always* – if or when there is a psychoanalytic process. Needless to say, this is a vague concept, too, and I have heard many different explications of it, depending, probably, on theoretical standpoints. Whatever the definition, it offers various kinds of circular reasoning. If such work as we choose to call "a psychoanalytic process" seems to have been going on, we may take any change that has taken place to be structural. And if there has been some change that we like to consider structural, we take this to mean that there has been a genuine psychoanalytic process. Contrariwise, if there has not been any structural-like change, we assume that no psychoanalytic process has been established, and if it is our impression that little or any genuinely psychoanalytic process has been going on in a clinically successful case, we conclude that whatever change there may have been has not been structural. You will realise, no doubt, that the essential questions about the structural change concept are indeed circumvented.

Some have tried instead to define the concept by listing the kinds of change that they think should be considered as structural. I believe that Dewald's contribution (1972) is the most thorough and will therefore cite it as an example. Dewald offered a vast number of examples of what he considers clinical indicators that structural change has been taking place, and he clustered them according to different phases in the psychoanalytic treatment. For instance, he suggests that, before the transference neurosis is well established, structural change shows itself by "how the patient applies himself to the task of psychoanalysis" (p. 309). By that he means that the psychoanalytic situation becomes established, that a therapeutic alliance is developing, that the ego is split in observing and experiencing parts, etc., listing further clinical inferences in this and later stages in the treatment. I carefully choose the word 'clinical inferences', because these examples are in most cases not empirical observations but distant theoretical inferences that would themselves have to be defined. The question that occurs to me is what is really contributed theoretically to these clinical inferences by invoking another and still more distant clinical inference. Also, we have to ask ourselves what it is about these changes that makes them structural and what it is about other kinds of change that will make them non-structural.

3. WHAT IS A STRUCTURE?

Let us begin our refinement of the concept by assuming that structural change has something to do with structures and that, if we want to develop a concept of structural change we have to begin with the concept of mental structure. The classical definition was offered by Rapaport (1967; Rapaport & Gill 1959), which was further developed by Applegarth (1989), thus: "A structure is often defined usefully and generally as 'something' which is the means by which a given function is carried out and which has a relatively slow rate of change compared to other processes... That is, it has a degree of permanence in psychological functioning. Under this definition, then, fall a wide group of processes with quite different levels of organization, degrees of permanence, and other varying qualities" (p. 1099). Carrying Applegarth's ideas a little further, I should like to define a mental structure as (1) a system or organisation, by which is meant that it is a set of elements that are interrelated in the sense of associating with, communicating with, or even influencing each other. Further, (2) this system is relatively stable or permanent and (3) does some kind of mental work, that is, has a function in the mental life or (4) organises some kind of mental content, that is, organises products of such functions. (5) It may be conscious or unconscious, and (6) it is not directly observable but will leave observable traces or derivatives in the form of human actions. (7) A mental structure regulates a mental function so as to bring about some degree of consistency in these actions. The structure will thus reveal itself by the consistency of its derivatives. Finally, (8) structures may themselves be organized in structures or systems, with all sorts of interrelations. Thus, there are structures superordinate to sets of subordinate structures, structures inhibiting or facilitating other structures, unconscious structures influencing conscious ones etc.

Now, this definition allows for a variety of all sorts of mental structures. Thus, we would call the different memory systems mental structures, because they perform the mental functions to lay down and hold memories. But, although they are products of these functions, we would not call single memories structures unless they were stable – which would rule out iconic memories and short-term memories. The internal object world is, on the other hand, a very extensive, long-term memory content, which is reflected in a certain consistency of feeling, thinking, and acting. At a lower level the specific internal objects are structures, too, again provided they are stable. Think of the super-ego as such a structure to which we have assigned special significance. In fact, we should conclude that any mental derivative that is observably recurrent, consistent, or predictable in human actions reflects an underlying, itself unobservable mental structure. Because not all human actions are recurrent, consistent, individually characteristic or predictable, it follows that not all actions are derivatives of mental structures.

The qualities of stability and consistency are central in the definition. Their relation is this: Whereas stability refers to the existence of a formation in continuous time, behavioural consistency refers to the agreement or correspondence among acts at different points in time, on different occasions or in different situations. Therefore, consistency presupposes a stable structure, whereas a stable structure will be expected to produce consistency. Consistency may thus be taken as an indication of an underlying structure, stable across the time taken to observe it, and the degree of consistency observed may be taken as a measure of the coherence of a structure, or its "structurality".

4. THREE KINDS OF STRUCTURAL CHANGE

But how do such structures change? I should like to differentiate between three kinds of structural change that I will show examples of later. I will first make a distinction between *structure* and *structuring*. Structuring is one way in which structures may change. It means, simply put, that something that was previously had less structure, that is, less stability and less of system-properties, has now more of them. Or, one might consider structuring as making a structure out of a non-structure, or stabilizing a system that is less stable. If, for a moment, we accept the energy metaphors of the economic point of view, an increased capacity of the ego to bind energy would be a psychoanalytic example of structuring, because binding denotes "an operation tending to restrict the free flow of excitations, to link ideas to one another and to constitute and maintain relatively stable [*sic*] forms" (Laplanche & Pontalis, 1980, p. 50). Clinically, we might see it manifested as increased impulse control, for instance. We may consider normal human development *partly* as a process of differentiation and integration in order to increase inner control and outer adaptation. Basically, this is a structuring process.

Secondly, I should like to distinguish between structuring and *restructuring*. A structural change in the form of a restructuring, simply put, is when there are changes in the layout, composition, form, or looks, of the structure. There are plenty of examples of intrapsychic restructuring: I guess the typical one, theoretically, is when a non-adaptive solution of a conflict is replaced by an adaptive solution, by rebalancing the poles of the conflict – even if this is only a metaphorical way of describing what is happening. Characterological change in the introjects, for instance as when the super-ego becomes more benign, or when the mother introject becomes less persecutory, are two other examples. In mental development restructuring is essential inasmuch as there are conflicts between new structures and earlier ones that have to be solved by restructuring.

Third, I should like to distinguish between structuring and *destructuring* and use the latter concept for a third kind of structural change. I believe you too have met persons who are a bit too structured, persons whom we regard as over-controlled, inhibited, compulsive, or simply too predictable. We sometimes want to destabilize their routine ways of being. So it comes rather natural to regard the obsessive-compulsive neurosis as an ideal case for destructuring. We should probably need a way to talk about these alternatives in other terms than diagnostic categories, however. Instead we may talk of opening up closed mental systems, of achieving an increased mental motility and openness, of turning rigidity in feeling, thinking, or acting into flexibility.

5. IS PSYCHOANALYSIS SUFFICIENT AND/OR NECESSARY FOR STRUCTURAL CHANGE?

The next issue is: Is the structural kind of change specific to psychoanalysis or may it come about in other ways and are all changes during a psychoanalysis structural?

Whether psychoanalysis is sufficient for structural change to occur is a question about the outcome of psychoanalytic treatments. Equating stable change and structural change, I guess there may have been at least one psychoanalytic case somewhere sometime that showed no such effect whatsoever, which would suffice to say that psychoanalysis is not sufficient. But that is a trivial argument: The decisive issue is whether psychoanalysis *may be expected* to produce stable change, that is, structural, change. That is a statistical inference. On the evidence collected in the Stockholm project, I would say, yes, it may. And definitely also a psychoanalytic therapy, provided it is not conducted in an as-if-psychoanalysis mode. And neither of these claims is trivial because, according to the limited number of psychotherapy outcome studies with long-term follow-up there are, some varieties of psychotherapy may *not* be so expected.

Now, to the question whether psychoanalysis is a necessary precondition for structural change. I shall start by pointing to some concepts outside psychoanalysis that have the qualities of structure. These are so-called dispositional concepts, like traits, attitudes, habits, and so on. For instance, a trait is a so-called intervening variable for whatever accounts for consistency in some aspect of individual behaviour, like conscientiousness or agreeableness. A trait would pass the stability criterion of a structure, also the unobservability criterion, and also the conscious-or-unconscious criterion. Perforce, a trait is a structure and a stable change in a trait has to be structural. So, there *are* structural concepts outside psychoanalysis, and there are also theories to account for changes in such phenomena, such as attitude change, learning, development etc.

In psychoanalysis, although there seems to be a general tacit assumption that structural change is a psychoanalytic specialty, many writers take it for granted that change of the kinds that we must regard as structural may occur outside analysis. This is a critical issue for my argument. Boesky (1988) says, "cures effected by manipulation of the transference or even by crude suggestion in some forms of psychotherapy also represent an alteration in compromise formation and therefore another kind of structural change." And he adds: "so do the spontaneous remissions of neurotic symptoms [...] as well as exacerbations of symptoms or the appearance of new symptoms" (p. 311). Abend (1990) exemplifies extra-psychoanalytical structural change with "identifications and responses to suggestion, or other forms of transference influence" (p. 532). We also have to consider the radical, profound, and stable effects that a traumatic experience will have on the psyche, whereas the normal, undisturbed spontaneous development must be regarded as a process of slow structural change. Considering the distinction between supportive and interpretative interventions, Kris (1993) suggested "that support can lead to enduring (structural [*sic*]) psychic change" (p. 112). Note this formulation: structural equals enduring.

6. ARE SYMPTOMS STRUCTURES, AND ARE SYMPTOM CHANGES STRUCTURAL?

Let us focus now on symptom change in this perspective. In their attacks and devaluation of behavioural therapies, many psychoanalysts have come to view symptom change as not only non-psychoanalytic but even anti-psychoanalytic, as if it were nothing but pseudo-change.

However, an interesting point of view on the status of symptoms was recently revealed to me in a discussion with a colleague of mine, a leading expert on cognitive-behavioural therapy internationally. His point was that "symptom" is a concept without a place in behaviour theory or behaviour therapy. His argument ran as follows. "Symptom" refers to a manifest sign of some underlying pathological structure or pathogenic process. However, behavioural theorists or therapists do not reckon with anything underlying at all. Manifest behaviour does not reflect or represent anything but itself and definitely not anything psychologically meaningful latent content. The manifest is the problem, simply. I think this converges with my argument that "symptom" is a term used to refer to indications or derivatives of underlying structures. Thus, I am *not* arguing that symptoms *are* structures. My position is that, provided that they are recurrent and consistent, symptoms are *derivatives of*, and therefore indicators of, underlying structures.

Reading psychoanalyst scholars on symptoms yields a somewhat confusing and contradictive picture. So let us rather look at what Freud said on this issue. He used the word "Ersatzbildung", "substitute-formation", and I believe the notion of a formation indeed implies a structure, or an arrangement or a disposition, as it is explained in Webster's Dictionary. In *Inhibitions, Symptoms and Anxiety* (Freud 1926d, p. 145) Freud suggests that this formation has two aspects – and both are structural in nature: "one, hidden from view, is the alteration in the id in virtue of which the ego is removed from danger; the other, presented openly, shows what has been created in place of the instinctual process that has been affected [by the defences] – namely, the substitutive formation." If these processes are structural, the reverse processes must be, too. Thus, symptom remission, as much as symptom formation, has to be considered indicating structural change, too.

7. THE ASSESSMENT OF STRUCTURAL CHANGE

Now, how should we assess structural change, clinically in outcome research? Let me summarize, to begin with. First, I have argued that there are all kinds of structures and therefore also all kinds of structural change, so this is not any psychoanalytical specialty. Second, I have argued that the essential quality of a structure is that it is relatively stable or permanent and that structural change itself has to assume a degree of stability or permanence in such stable or permanent dispositions. Third, I have argued that structures, and thereby structural change, are themselves unobservable and therefore *have to be inferred on the basis of indications or indicators*. This is exactly equal for the clinical situation as for the research situation. Fourth, I have argued that stability in underlying structures, given enough time, will produce consistency in its derivative actions.

In consequence of these arguments, we cannot define structure and structural change by pointing to some psychoanalytical function or functions and claim that these, but not others, are structural, and consequently there are no particular measures or scales or instruments that, as such, specifically address structures or structural change. This is not to deny that some functions may be more central to what a psychoanalyst is trying to effectuate, be they ego functions, object relations, drive processes or what not. But the structurality of a function does not really reside in the function and therefore nor in any scale measuring it. Stability and consistency, which are the *definiens* of structurality, do only reside in the application of the scale in the assessment design.

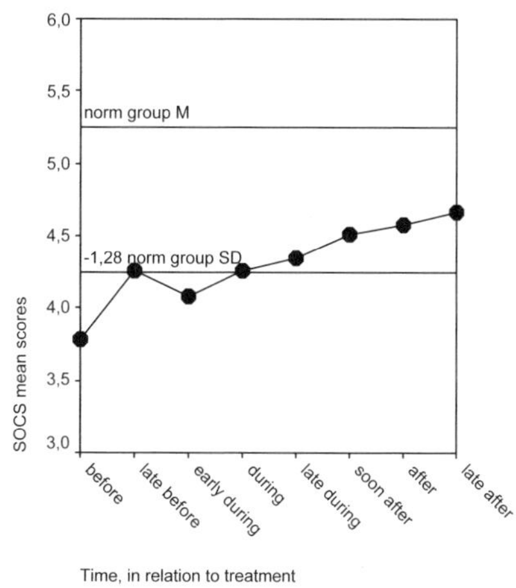

Figure 1. Sense of coherence

Now, there are indeed a great number of instruments *claiming* to study *consistency*. Indeed, the typical method in personality psychology is focussing on inferred traits by the use of so-called personality inventories for self-report or rating scales for observers. The interesting thing about these methods is that they do not really *assess* consistency, in the sense of actually making repeated observations across time. Rather, consistency is typically taken for granted in the questions or the instructions. What they typically do is to ask for generalizations across time or across occasions, like “How typical is this for you or him or her?” or “How often do you or he or she do this?” etc. Obviously, consistency – and thereby stability – is assumed, only. But, equally obvious, any person’s response to an item of that type will be more or less influenced by temporary conditions inside or outside, and that is of course why Mischel (1968) long ago noted that the validity of personality inventories in relation to action were quite modest. It is also likely that an unknown share of consistency reflects so-called response sets that are independent of the disposition one was really aiming at, such as yea-saying, nay-saying, social desirability. So, although they are convenient to use, the validity of these instruments is based on generally untested assumptions. Consider the phenomenon called the sense of coherence. It is claimed, by its originator, Aaron Antonovsky, to be a very stable and enduring, almost unchangeable, information-processing system. Look at Figure 1 (page 273) what happened with the sense of coherence during psychoanalytic treatment in the Stockholm project.

Obviously, it is *not* stable under all conditions. On the contrary, it seems to change all the time, most of the time upwards, to the better, sometimes, in the beginning of treatment, downwards, to the worse. Consequently a single administration of Antonovsky's questionnaire would not be enough to infer that we are measuring a structured function.

So we have to use a stronger assessment design, the repeated-indications design. It makes time an accessory, by having repeated indications and inferring structure by observing the consistency among these repeated indications. And note that, if these are numerous enough, there may be some intervention interspersed between them, which makes the inference of structure one of structural change, really. Although this design is as relevant to clinical in-session and between-sessions observations, for systematic evaluation or research purposes, we may of course use the same instruments as above, self-rating personality inventories, observer rating scales or whatever; the important difference is that in this design we make repeated observations, *exploring* the consistency of the phenomenon we are interested in rather than taking it for granted. Assuming that stabilisation does not depend on the clinician or the instrument being insensitive to change, or that the patients respond to on the basis of response sets such as social desirability, that is, to please the analyst or therapist, we may draw safer conclusions about what amount and kind of structural change there has been. This is actually what we did in the Stockholm project. The details about how this study was designed are not really necessary for the present discussion, so suffice it to say that we used a number of overlapping groups of cases in psychoanalysis or long-term psychoanalytic psychotherapy, 434 cases altogether, to reconstruct a general trajectory of outcome across an ordinal time scale, ranging from two years before treatment to three years after termination. Figures 2 and 3 show the development across time in symptom distress and social malfunctioning. Note that in these two figures, improvement is a downward trend, in contrast to the previous figure.

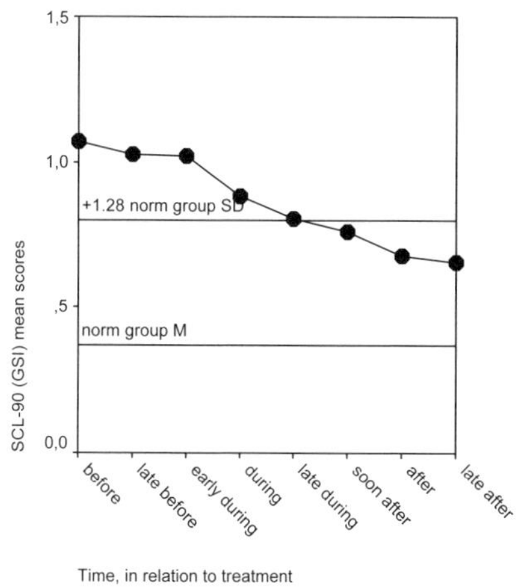

Figure 2. Symptom distress

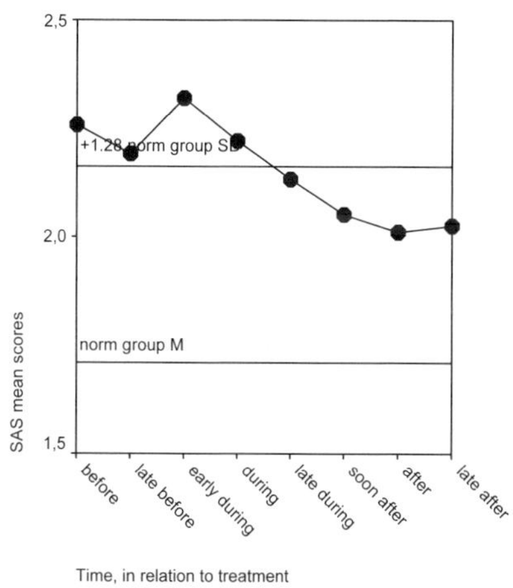

Figure 3. Social malfunctioning

Now, is it reasonable to infer any structural change from these trajectories? It seems so, it seems as if the patients, at an average, were in a steady state before treatment and found a new and improved steady state a couple of years after treatment that is compatible with what would be expected of structural change, at least when social relations and symptoms were concerned. And, undoubtedly, it is a lot easier to believe in stable or structural change there than if we had merely measured these variables once before treatment and once after, of course. But, if we now focus on the *post*-treatment period and decompose the entire sample depending on treatment duration and session frequency, we find that development post-treatment may look very different. Figure 4 is for the sense of coherence again, and the others look very much the same. Depending on duration and session frequency, post-treatment change may be more or less positive (upward trends) but under some conditions even negative (downward trend).

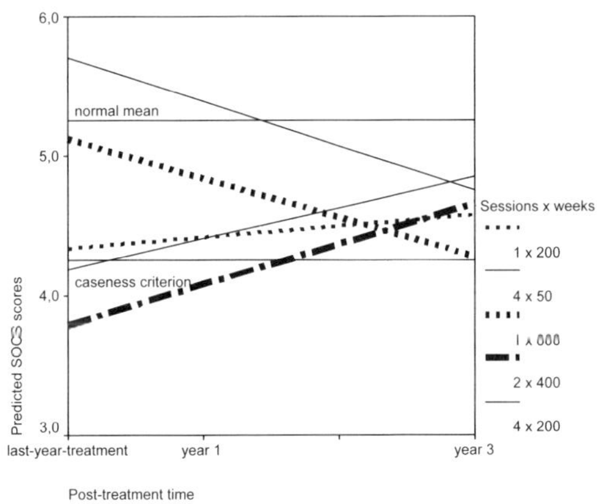

Figure 4. Sense of coherence (post-treatment)

These are so-called modelled data, which is why they are so clean and regular. But where is structural change here? A steady state means that there is no change at all, but we should presumably prefer a not-steady state in the sense of continuing improvement, like the thick solid line!

A weaker version of the repeated-indications design is to consider the various items of the instrument one uses as repeated indications. This only weakly tests the *stability* criterion but may be used as an approximation, because it nevertheless indicates *consistency*, in a sense. Thus, if a function has stabilized, we would expect the patient to respond in similar or corresponding ways to different tests of this particular function. Using a standard instrument, as we did in the STOPPP, we should expect that different items would settle on roughly the same level, thus having *a smaller dispersion or variation among themselves* later than they had before. Look in Figure 5 what happens in psychotherapy and psychoanalysis, when the sense of coherence is concerned: A gradual decrease in such dispersion and a final reduction after psychoanalysis, by about 50%, visibly less after psychotherapy.

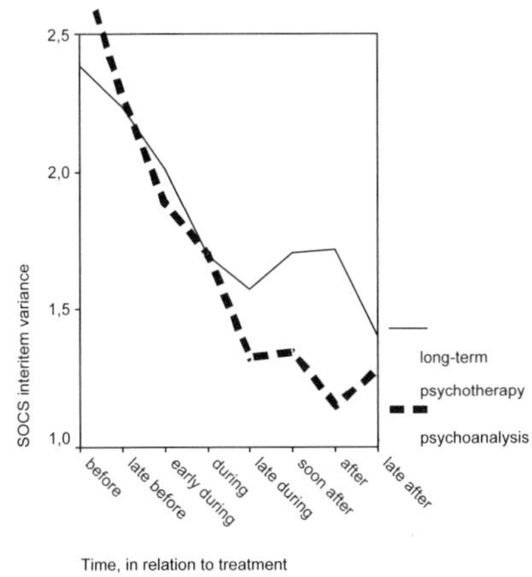

Figure 5. Sense of coherence. Long-term psychotherapy and psychoanalysis

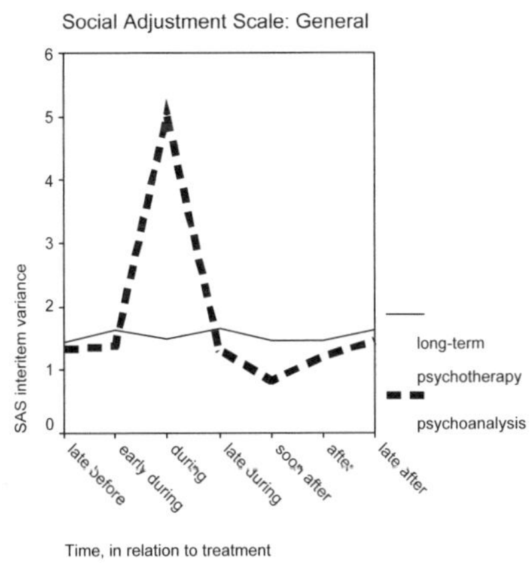

Figure 6. Social Adjustment Scale: General

The same picture applies to the symptom distress measure. But Figure 6 shows what happened when social relations were concerned. A spike in the middle of the psychoanalyses and no change at all in psychotherapy! Although this does *not* look like a case of structural change, because it is transient, not stable, it has to be explored further. We may decompose the social relations scores in a number of subscores. Most of them exhibit a steady structuring, as in Figure 5. But in Figure 7 we find another one, concerning one's relations to friends, with exactly that upward spike during treatment.

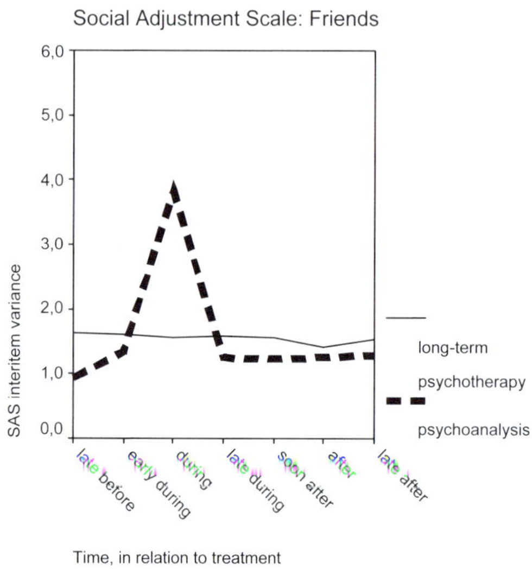

Figure 7. Social Adjustment Sacle: Friends

Finally, let us look at agreement, not across items but across time. As all patients in the study delivered data three consecutive years, we may explore the consistency across these years. What we see in Figure 8, when social relations are concerned, is a sharp drop in consistency from year to year early in treatment from a *very* high initial level and then a gradual increase to a lower level than initially. In the meantime, we also see a sharp drop that is the inversion of the spike we saw earlier. We may interpret the initial reduction in correlations from year to year as a kind of destabilization early in treatment, exactly what I had in mind talking about destructuring. It is structural, because the correlations do not revert to their initial, high level. So it seems that psychoanalytic therapy appears to provoke revisions of one's priorities in one's social relations. What happens after that may be interpreted as structuring, on a more flexible level.

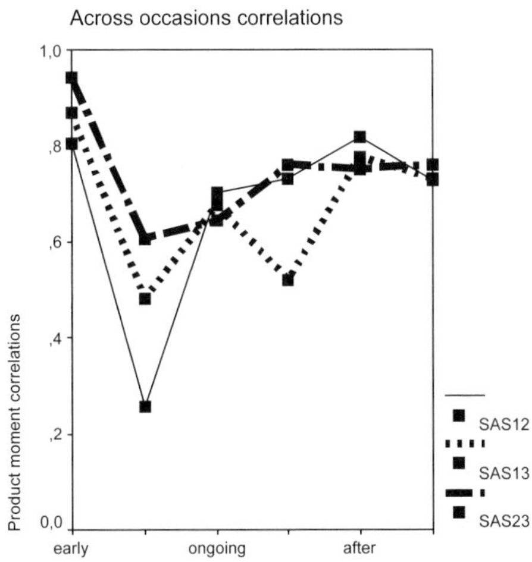

Figure 8. Across occasions correlations

Now, in conclusion, are these indications of *structural change*? I think they are, like any *stable* change in any *stable* mental function.

REFERENCES

Abend, S. (1990). The psychoanalytic process: Motives and obstacles in the search for clarification. *Psychoanalytic Quarterly* 59: 532–549.

Applegarth, A. (1989). On structures. *Journal of the American Psychoanalytic Association* 37: 1097–1107.

Blatt, S.J., Chevron, E.S., Quinlan, D.M., Schaffer, C.E. & Wein, S. (1988). The assessment of qualitative and structural dimensions of object representations (revised edition). Unpublished manual. New Haven, Yale University.

Boesky, D. (1988). Comments on the structural theory of technique. *International Journal of Psycho-Analysis* 69: 303–16.

Dewald, P. (1972). The clinical assessment of structural change. *Journal of the American Psychoanalytic Association* 20: 302–324.

Freud, S. (1926d). *Inhibitions, Symptoms and Anxiety. S. E.* 20: 87–172.

Freud, S. (1974). *Indexes and Bibliographies. The Standard Edition of the Complete Psychological Works of Sigmund Freud*, ed. J. Strachey. Vol. 24. London: The Hogarth Press.

Kris, A.O. (1993). Support and psychic structural change. In *Psychic Structure and Psychic Change*, ed. M.J. Horowitz, O.F. Kernberg & E.M. Weinshel. Madison, CT: International Universities Press, pp. 95–115.

Laplanche, J. & Pontalis, J.-B. (1980). *The Language of Psycho-Analysis.* London: Hogarth Press.

Mischel, W. (1968). *Personality and Assessment.* New York: Wiley.

Rapaport, D (1957). Cognitive structures. In *Contemporary Approaches to Cognition*, ed. J. S. Bruner et al. Cambridge, MA: Harvard University Press, pp. 157–200.

Rapaport, D. & Gill, M. (1959). The points of view and assumptions of metapsychology. In *Collected Papers of David Rapaport*. New York: Basic Books, 1967, pp. 795–811.

Strachey, J. (ed.) (1974). *Indexes and Bibliographies. The Standard Edition of the Complete Psychological Works of Sigmund Freud.* Vol. 24. London: The Hogarth Press.

Weinshel, E. (1990). Further observations on the psychoanalytic process. *Psychoanalytic Quarterly* 59: 629–649.

XII. ON THE CONTRIBUTION OF THE MOST RECENT RESEARCH ON LANGUAGE APPLIED TO PSYCHOANALYTIC PRACTICE

JORGE CANESTRI

The fact that analytical practice, named the "talking cure", as well as the theory on which it is based, has to come to terms with language and the various disciplines that deal with it, may appear to be a truism. However, in psychoanalysis this confrontation is not treated with the necessary solicitude or awareness. Even less frequently do the linguistic disciplines take an interest in the psychoanalytical insights on the nature and functioning of the psychic apparatus, or consider the exquisite peculiarity of clinical experience as a possible area of research.

The objective of this brief paper, in which I will try to take some regional differences into account, is to give an overview of the problems that recent research in the vast field of work on language put to psychoanalysis. It would obviously be possible to reverse the perspective and consider the problems identified as being an input to research methodology in the disciplines mentioned above. I should like to clarify that I am speaking about research on language and not about linguistics, because in this research linguistics are certainly present, but so are the neurosciences, psycholinguistics, philosophy of language, semiotics, etc.

The relationship between psychoanalysis and research on language may be approached by referring to the psychoanalytical works that have attempted to present a critical, theoretical and historical survey. This also allows us an intuition into the regional differences. In general, the object of these works is mainly the relationship between psychoanalysis and the theory of linguistics in the strict sense. I will mention three works in particular from three different regions (North America, Latin America and Europe) because they clearly illustrate the differences in formulating the problem.

The first is the well known paper by B.E. Litowitz and N. Litowitz (1977) published in the *Journal of the American Psychoanalytic Association*, "The influence of linguistic theory on psychoanalysis. A critical, historical survey". The title reveals that in this work also the direction of the reading is from linguistics to psychoanalysis and not *vice versa*. The authors are aware of this when they write: "In addressing the topic of the influences of psychoanalysis as an intellectual system upon linguistics as the systematic study of language we are viewing a subject with little, if any, past, some present and an open, hopeful future" (p. 419). Undoubtedly, this is one of the clearest and most complete works as far as formulation, identification of problems, and linguistic competence are concerned. In it are identified "three major shifts in linguistic theory (labeled nineteenth-century, modern and recent)" and an attempt is made to explain the reasons and advantages for each shift. The authors inform us that "Following each description, psychoanalysts will be grouped as having been influenced by one or another of the major linguistic theories" (p. 419). The list of the authors and their corresponding works is very complete regarding English-speaking psychoanalysts, and it surprisingly also includes a very relevant author from Latin America who has not been translated into English and who belongs to the last shift described in the paper (recent, Chomskian-oriented linguistics): I am talking about David Liberman (1983). Among the French-speaking psychoanalysts the "linguistic" concepts of J. Lacan are carefully analyzed and A. Green is mentioned, but not his works that are more specifically dedicated to the relationship between language and psychoanalysis. There is no reference to German-speaking authors except for the philosopher Habermas. I mention these aspects about which thinkers are taken into consideration and which are not, because it helps to understand thc rcgional differences. The work mentioned above was written in 1977, and although it is one of the most complete, it is certainly outdated. "Recent" linguistics goes as far as the most classical theories of Chomskian generative grammar, already abandoned by the American linguist, while the current generative linguistics is dominated by two strands: a) the Theory of "Principles and Parameters" (*Knowledge of Language* 1986) and "Minimalism" (*The Minimalist Program*, 1995).

If, on the other hand, we consider the work of authors from Latin America, besides the use of important linguistic theories we also note in them a careful examination of the bibliography relative to the relationship between psychoanalysis and language in other idioms (English, French and German), but with little interest in the North American psychoanalytical bibliography on this theme. I quote a paper by Susana Vinocur de Fischbein (1998): "Psicoanálisis y lingüística: contactos e intercambios" (Psychoanalysis and linguistics: contacts and interchanges), that, besides offering a review of the relationship between linguistics and psychoanalysis and a summary of the Latin American position on the matter, also inquires into present day linguistics and suggests some possible areas of research. The author underlines the fact that linguistics has shown its influence on three different areas of analytical theory and practice: a) one that could be called 'psychoanalytical psycholinguistics', comprising the theory of symbol formation, of the symbolic processes in general, and of the acquisition of language; b) another that attempts to individuate the role of language in the structurization of psychism: language as a function and/or psychic structure, c) the third is the area regarding the meaning of language in clinical practice. She analyzes some of the most interesting aspects of contemporary research on language, for instance pragmatics, intertextuality, inferential pragmatics and its theory of implicature (Grice) or of relevance (Sperber and Wilson) and the discourse analysis. From this point of view, her work, although in terms of historical and critical aspects considerably less exhaustive than the previously mentioned work, is much more in agreement with the most recent research.

Turning now to the European region, the choice of a representative work becomes more complex: firstly because there are no summarizing papers similar that of Litowitz; and secondly because they are strongly characterized by theoretical standpoints that are usually clearly defined and even polemical. The panorama ranges from Lacan's 'linguisterie' to Lorenzer's hermeneutics. Perhaps the most relevant aspect of the European formulation of the linguistics-psychoanalysis relationship is the fact that the emphasis is always on psychoanalytical metapsychology, in other words on the requisites that a linguistic theory must have in order to be compatible with psychoanalysis. If this is not so, then the linguistic theory can be deformed (Lacan's *linguisterie*) or not taken into consideration.

A book that I could quote is by the French linguist and psychoanalyst L. Danon-Boileau, *Le sujet de l'énonciation* (1987), in which the author outlines the field of linguistics-psychoanalysis interaction following the classical divisions of the Freudian model: economic, dynamic, topical and 'genetic'. In this framework every psychoanalyst (mainly French and some well-known English analysts such as Winnicott, Bion and Segal) finds a place. Litowitz did something similar when she compiled a classification of the authors according to the influence that the linguistic theories had had on them.

Another significant text seen from a very different viewpoint is the well-known "Le langage dans la psychanalyse", by A. Green (1983). In it, as well as in a subsequent work "Le langage au sein d'une théorie générale de la représentation" (1997), he sets out the conditions that a theory on language must comply with in order to be acceptable to psychoanalysis. Perhaps, as we shall see later, Green's major contribution to the linguistics-psychoanalysis relationship is the general theory of representation.

Although looking at the issue of the relationship between psychoanalysis and research on language from the point of view of the above mentioned works can give a rough idea of the differences between the regions, it does not offer an adequate outlook on the problem in general. It could be better defined through an accurate analysis of some of the problems that linguistic models present when the analytical experience has to be taken into account. A minimum outline of the relevant concepts in the field of research on language that might weigh on psychoanalytical practice and theory should include:

1. *Philosophy of language:* the change of the classical, propositional paradigm (semantic-logical-referential) with a pragmatic, dialogical one. Pragmatics and its importance as performance theory at the semantic level. The inferential pragmatics and its theory of implicature (Grice) or of relevance (Sperber and Wilson) and its influence in clinical practice.

2. *Discourse analysis.*

3. *Rules of transformation in listening and interpretation.*

4. *Acquisition of language.* Biological bases of language, neural plasticity, research in neurosciences. Early interactions, pathological dyads. Inner language, language of thought, language and thought.

5. *General theory of representation.*

I will very briefly try to illustrate these issues:

1. PHILOSOPHY OF LANGUAGE

To explain the meaning of the change of the classical paradigm in a new one I have used, in a chapter of the book *Changing Ideas in a Changing World* (Canestri 2000), an anecdote that the German philosopher Karl Otto Apel proposes at the beginning of his book *Le Logos propre au langage humain* (1986). I repeat it here because I think it clearly illustrates the problem of the two paradigms of the philosophy of language. Karl Otto Apel recalls that his reflections were suggested to him by a comment of Karl Popper. Following a lecture of Apel's in which the philosopher attempted to outline the program of a pragmatic-transcendental theory of communication, Popper said that it is not necessary to pay so much attention to communication inasmuch as it is what we have in common with animals, and that what is important in human language are the propositions. This comment sums up the problem perfectly. Popper very precisely defined the classical paradigm that from Aristotle to Frege informs the philosophy of language. "The classical or propositional paradigm to which Popper refers, is the semantic-logical-referential of the *logos*. As Apel reminds us, the platonic root of logos, on the other hand, was dialogic: logos as language and discourse in conversation. From the viewpoint of the classical paradigm there would be a division and an opposition between referential semantics and simple pragmatics of the communicative functions of language. The classical paradigm does not consider that pragmatics contributes to the semantic logos, to the inter-subjectively valid meaning of the signs of language and to the meaning of theories. A conception of linguistic activity that does not recognize the contribution of pragmatics to the construction of the meaning, that is, its semantic contribution, would be extremely limiting in analytic practice and would contradict our experience" (Canestri 2000, pp. 198–9).

Probably no psychoanalyst in his clinical practice has ever neglected this aspect relative to the contribution of pragmatics in the construction of the meaning, inasmuch as Freud, in the third chapter of his rightly famous book *Jokes and their Relation to the Unconscious*, proposes the well known joke: "Two Jews met in a railway carriage at a station in Galicia. 'Where are you going?' asked one. 'To Cracow', was the answer. 'What a liar you are!' broke out the other. 'If you say you're going to Cracow, you want me to believe you're going to Lemberg. But I know that in fact you're going to Cracow. So why are you lying to me?'" The following comment by Freud anticipates several of the above mentioned issues. He writes: "But the more serious substance of the joke is the problem of what determines the truth [...] Is it the truth if we describe things as they are without troubling to consider how our hearer will understand what we say? Or is this only jesuitical truth, and does not genuine truth consist in taking the hearer into account and giving him a faithful picture of our own knowledge" (1905c, p. 115). This excellent example certainly gives us food for thought, as Freud himself notes. But what I wish to emphasize in this case is the fact that Freud places pragmatics in the foreground; in fact, in the communication between the two Jews, the truth does not depend on the description of "things as they are", but on what the subjects do with the words (Canestri 1999a).

Certainly, applying an implicit knowledge of the function of pragmatics in the construction of the meaning in clinical practice, is not the same as working and carrying out systematic research in this perspective. In the three regions mentioned there are some analytical works that take these principles into consideration. I can mention the already quoted S. Vinocur de Fischbein (1998) in Argentina, T. Shapiro (1988, 1991) in the U.S.A., L. Danon-Boileau (1987) and myself in Europe, and I could list many other names from every region.

2. DISCOURSE ANALYSIS[1]

Putting the accent on the linguistic production of the individual rather than on the ideality of the language, favours the growth of an area of research entitled "analysis of the discourse", which privileges the structure of spoken language as manifested in current use: conversation, interviews, etc. The 'discourse' is opposed to the 'langue' (language, because we have no other word in English to name the French opposition between 'langue' and 'parole') since it implies the use of the 'langue' in a certain context. As Charaudeau and Maingueneau (2002) point out, the proliferation of the term discourse in the sciences of language "is a symptom of a modification in the manner of conceiving language" (p. 187). Language is studied as a dynamic, social, intersubjective phenomenon, intrinsically dialogic, in the original meaning of platonic "logos" (mentioned before) or in the more recent meaning resulting from the studies of Mikhail M. Bakhtin. It is possible to make a typology of discourse, and to describe some of its defining characteristics: to be a form of action, to be contextualized and interactive, etc. A first field of research for psychoanalysis would be the description and careful study of the specific characteristics of the psychoanalytic discourse (which could be conceived as a particular type of discourse) that is so profoundly different in rules, structure and codification from any other discourse of ordinary communication. To offer a few examples: the endofasic internal discourse of patient and analyst, that elsewhere I have called "internal flux" (Canestri 1994a), does not have to take pragmatics into account, but pragmatics reappears in the uttered sentences of both. Nevertheless, the existence of the fundamental rule of free association and free floating attention allows both participants of the experience to transgress canonical rules in the analytic communication. In analysis the word expresses the drive and is also a drive object that must be interpreted (Fónagy 1983); there are a great many envel-

[1] This paragraph is a re-elaboration of some of my previous works (Canestri 1999a, 2000).

oped registers (preverbal, organic, paralinguistic), conditioned in their turn by the context (internal transcriptions, transmutations of code, etc).

The analysis of the discourse brings to the forefront the production of utterances. The different forms that the presence of this act of production can take, and also the degrees of its importance, enable a specific typology of the discourses to be created. From the point of view of the analytic discourse, the study of the elements of the utterance that allude to its production is particularly significant because they are part of the actual sense of the utterance. Some of these allusions are represented by grammatical categories such as personal or demonstrative pronouns, relative adverbs and adjectives or tenses of verbs (i.e. deictic categories); other allusions are represented by semantic categories such as the time and the place of the utterance.

In the follow-up of the analytic process we can obtain a great amount of knowledge about psychic change by analyzing the discourse of the patient. This implies considering to whom the patient speaks (transference) and how he perceives himself, not only through the analysis of the content of his unconscious fantasies. It is important also to observe the linguistic means that he uses (and that are normally unknown to the speaker), e.g. personal pronouns, to represent all this in the discourse. Within this vast field of research I should mention the work of the psychoanalyst and linguist L. Danon-Boileau (1987) (France) in that, as a disciple of Antoine Culioli, he proposes a "linguistics of the utterance" that he considers could be particularly useful in the analysis of the psychoanalytic discourse. Certainly, many others could be quoted, and with many different theoretical orientations ranging from those who do simple computer-assisted analysis of lexical frequency to those who, like E. Mergenthaler and W. Bucci propose an interpretation of the referential process in the psychotherapeutic discourse context by computer assisted text analysis (1998).

3. RULES OF TRANSFORMATION IN LISTENING AND INTERPRETATION[2]

In the linguistic activity of individuals, language becomes a complex and heterogeneous object. All the more so in the analytic discourse where the three classical channels described by semiotics are co-present

[2] In this paragraph I re-elaborate some ideas I discussed in previous works (1994a, 1994b, 1999a, 2000).

and cannot be dis-articulated. I refer to the first channel of articulation (segmental) that includes, in order of increasing complexity, phonology, morphology, syntax and stylistics; a second channel (or supersegmental level) that includes a wide range of phenomena coming under the classification of prosody (tone, accents, pauses, silences, intonation and expressive connotations); and a third channel or level including non-verbal elements such as gestures, facial mimicry, movements and positions of the body, etc. If we can say that the activity of the analyst relies to a large extent on his/her capacity to contain and adequately transform the unbearable experiences of the patient through a correct de-codification of the patient's communications then the importance of a conception that takes into account the complexity of the individual's linguistic activity and of the whole range of semiotic systems both for production and for recognition, becomes all the more evident.

It is necessary to carefully reflect on the meaning of a process that is frequently formulated as an identity: "to decode, that is, to interpret" and also on the nature of the transformational rules in it. Keeping within the area that I have tried to define, I think that the work of H. P. Grice (1989) can be of assistance to us. Grice formed part of the Austin group, and after the rejection of neo-positivism that lead the "Oxford Philosophy" to abandon all types of formalization, he was the one who, in the field of the philosophy of language, was able to develop theories. Here I should like to mention just four very important distinctions introduced by Grice: (a) between the natural meaning and the non-natural meaning; (b) between what the words mean and what the speakers mean; (c) between what is "said" and what is only "implied", and (d) between the conventional meaning and the non-conventional meaning.

Grice tries to explain what a single speaker in a given circumstance means by a word or a gesture. It is this concept of meaning, that Grice calls "the meaning of the speaker", that precedes the concept of the linguistic meaning and not vice-versa. It is not the conceptof conventionality (distinction between natural signs and conventional signs) that distinguishes the non-natural meaning from the natural meaning, but the concept of intentionality (connection between meaning and intentions, between semantics and psychology). If I mean something, I must intend to obtain a given effect on my interlocutor, at least in part through his recognizing my intentions. This is where a vast field of possibilities opens up from the psychoanalytical viewpoint. The intentionality that concerns us as psychoanalysts is not the conscious one; the effect that a patient tries to obtain does not correspond to a conscious but to an unconscious wish (and a good example could be the way projective identification operates). In Grice's theory the meaning of a sentence ultimately depends on what the speakers mean when they utter it, as long as the linguistic use is correct. When we converse, we often mean more than what the words mean literally, or we mean something different from their literal meaning. Grice here introduced the concept of "implicature"; the "implicature" has the task of systematically linking what is said to what is implied, therefore explaining how it is possible to do what we do continually when speaking, i.e. communicating more than our words mean literally. A corollary of this statement is that a proposition is certainly conveyed by an utterance, but this depends not only on the utterance pronounced, but also on the circumstances of the utterance and on a series of general rules governing the conversation and the use of the language.

Comparing Grices' ideas with those of Davidson (1984) on similar issues is very interesting. Davidson postulates that in order to understand the meaning of an utterance it is necessary to insert the discourse into the general framework of rational action. He emphasizes that an utterance directed to an interlocutor is an action with two components: the wish to produce a reaction depending on the semantic characteristics of the utterance, and the belief that the interlocutor has the necessary resources to understand the utterance just as we ourselves intend it. Moreover, Davidson considers that during the course of the exchange we provide the other with a series of clues that help to orientate him.

Another step forward has been made by Sperber and Wilson (1986). The starting point of these authors is a reflection on the classic model that tries to explain how communication takes place. According to the classic model or *code model*, coding and decoding messages is communicating. Grice's theories (but also Davidson) suggest a different model, the *inferential model*, where communication consists of the production and interpretation of facts. These two models are not incompatible and can be combined in various ways. Verbal communication brings into play at the same time both inferential processes and codifying processes (as can be seen in Davidson's conceptualisation). A code is a system that associates messages with signals and requires a codifier and a de-codifier. The inferential model must include central processes at the entrance and at the exit, a set similar to: thought processes→acoustic codified signal→acoustic signal received and de-codified→thought processes. From this point of view the semantic representation of a sentence does not coincide with the thoughts that can be communicated by pronouncing the sentence. We pass from semantic representation to communicated thought not through an addition to the code but by means of inferentials.

It is clear that the distinction between sentence and utterance is essential; an utterance possesses a set of characteristics, some of them are linguistic, others not. The study of the interpretation of utterances belongs to pragmatics, utterances express propositional attitudes, they carry out linguistic actions, or else they have "illocutionary force".

Summarizing, from this viewpoint, communication succeeds not only when the speakers recognize the linguistic meaning of the utterance, but also, and sometimes predominantly, when they infer the "meaning" of the utterer. It is worth reflecting on how and when this operates in psychoanalytic practice.

From the point of view previously illustrated (Grice, Davidson, Sperber & Wilson) the patient as well as the analyst both "say more" than they are aware of. This is normal on the part of the patient and is, in a way, at the basis of the possibilities of our inferring and interpreting. Generally we are less aware that our interpretations also "say more" than they apparently do not say. After all, we are not always aware of the mechanisms that moderate our comprehension of the patients' material, and of how much of that comprehension we remain unaware of and yet is transported in the interpretation, not through its content but through its implications. Equally often the therapeutic quality of an intervention cannot be measured from the content itself of the interpretation, but from the implications that it contains. It is like saying that "we are doing well" without knowing it, i.e., that the "doing well" is frequently a quality in the analysts' unconscious that is explicated through his or her speech. This argument is closely linked to the discussion about the "therapeutic factors" in analysis. For lack of space I am unable to propose some clinical examples that would illustrate the type of deconstruction of the analytical dialogue that these concepts make possible.[3]

4. ACQUISITION OF LANGUAGE

We cannot discuss language – and even less so the relationships between sciences of language and psychoanalysis – without considering how language is acquired. Nor can we hide the fact that there is no ex-

[3] I proposed one in Canestri (1999b).

planatory monolithic theory, while it can be said that the new techniques that study the emergency and consolidation of language have reached a high level of refinement and reliability, in many cases revolutionizing the knowledge that we possess.

I only want in this paper to underline those aspects that I think are linked to the psychoanalytical experience and that psychoanalysis must take into consideration. Probably it is the knowledge acquired through the use of new techniques that permits us to trace a new and perhaps median route between the two great families of theories that deal with the explanation of how children come to acquire language. The first of these two families privileges those linguistic constrictions that are considered innate (learnability theory, theory of Principles and Parameters, theories of bootstrapping); the second emphasizes the cognitive and communicative determinants of acquisition.

Each of these families, as they evolved in order to formulate theories more adherent to the data provided by experience, has in its turn created solutions including elements from the other family. Let us for example consider the evolution of the family of theories that privileges innateness. From the initial Chomskian position on the existence of a Language Acquisition Device, Chomsky himself moves on to the theory of Principles and Parameters that gives rise to the 'learnability theory' and to the idea that a suitable environment is only a trigger for the innate linguistic faculty (principle of 'I language'). However, also within this current of thought, this extreme position is modulated by the theories of 'bootstrapping' (syntactic, semantic and prosodic bootstrapping); Pinker (1987), for example, favors the intervention of non-linguistic capacities (perceptive and cognitive) that contribute to the acquisition. It is, however, fairly evident that the theories of this family do not take into account the psychological mechanisms that might intervene, or the factors linked to development, or the role of the environment beyond serving as a trigger. From the point of view of psychoanalysis, these theories, at least in their most radical versions, prove to be unsatisfactory since they do not answer any of the problems raised by clinical experience, also with very young children. They are just as unsatisfactory when considering childhood language pathologies. The experience of those specialists – neuropsychologists and child neuropsychiatrists – who treat them contradicts many of their basic hypotheses and is more in agreement with a neo-Piagetian type of cognitive hypotheses.

Communicative and cognitive theories on the acquisition of language present, in the same way as the previous ones, many versions and variations. The theory called interactionist, based on the linguistic theory of Austin and Searle of the *Speech Acts*, does not attribute to the child an initial language endowment, but it assigns to social exchanges the task of bringing the child to the construction of abstract linguistic units, of grammatical rules, etc. I think this version, that appears to be just as radical as the most radical innatist positions, is of particular interest concerning the pragmatic dimensions of language that I dealt with previously. It allows us to explore a dimension that is essential from the point of view of psychoanalysis and that is overlooked by other theories, i.e. the dimension of the effective and affective mother-environment/child exchange and the constitution of more or less normal or pathological dyads. I shall return to this issue later on. However, it is clear that from the point of view of language acquisition as such, this theory leaves many questions unanswered.

The cognitive theories underline the need to link language and cognition and to thus hypothesize the existence of cognitive prerequisites in order for language development to be possible. There are many authors representing many modulations of this principle, but I mention only A. Gopnik (1997, 2000) and Karmiloff-Smith (1992). It is fairly evident that these theories are related to Piagetian cognitive models, although they attribute to the child many more inferential capacities than Piaget conceded to them; besides, this is in line with the development undergone by the School of Geneva. Research in this area has also stimulated the cross-linguistic studies of language acquisition and these have proved to be fundamental from many points of view.

Among the cognitive theories according to which language acquisition derives from a general cognitive base, perhaps the most interesting are those that use connectionist perspectives on development (the title of a book written in 1996 by Elman, Bates, Johnson, Karmiloff-Smith, Parisi and Plunkett) and place the accent on the non-linear development of auto-organized systems with successive modularization. To my mind, this perspective is interesting because it is nearer to certain facts evidenced by neuro-scientific research. Probably the moment has come when this research, made possible, as I have said, by the new exploratory techniques (essentially neuroimaging techniques), will help to construct a third family of theories more radicated in the data coming from our present knowledge about the biological bases of language and about cerebral organization, its plasticity and functional specialization relative to development. In the same sense, we must pay attention to the re-educative experiences of infantile language pathologies because they can provide data of great interest. The perspectives of this type of research are innumerable and the rhythm of the discoveries is in continual acceleration. To quote one example, in the issue of 18 October 2002 of *Science* an interdisciplinary and international team working at the Sissa in Trieste (Peña, Bonatti, Nespor & Mehler 2002), using precise experimental methods, have shown how, very early on, a baby can distinguish in the continuous flow of spoken language (also when the researchers used an invented language) the end of one word from the beginning of the next. This is based on the possibility to make unconscious mental calculations that reveal statistical regularity, which is a computational aspect of the brain.

Many of the questions we have asked ourselves about the different aspects of language acquisition including, for example, very early bilingualism and the cerebral organization of two or more languages, today find convincing answers in the techniques of neuroimaging. Many of the hypotheses we formulated in the past before we had these techniques can now be confirmed or rejected with a high degree of certainty. For example, in the work carried out together with colleagues J. Amati Mehler and S. Argentieri (1993) on bi- or polylinguism in analytic practice, our theory on the hypothetical configuration of languages in the brain and on memory organization has been partly overtaken by more recent research using neuroimaging (cf. Fabbro 1999). Without this research, the mechanisms in question would remain unknown.

As research stands at the moment, it is reasonable to think that, in the cerebral organization of multilinguals, the representation of the mother tongue occurs through ways and processes that partially differ from those used for the representation of other subsequent languages. Within the mosaic of the memory systems, the acquisition of the mother tongue seems to be linked to the procedural memory, i.e. to an implicit mechanism that does not require awareness and that functions through repetition of the task, with the involvement of the sub-cortical structures connected to the prefrontal associative areas. Even if there is a partial overlapping of the areas involved in the use of languages that have been acquired during different periods of life, some anatomic and functional differences are noticeable in the group of "late bi-polylinguists" according to the language used from time to time (Paradis 1995; Fabbro 1999; Kim, Relkin, Lee & Hirsch 1997).

Besides my clinical experience, the proved existence of dynamic rearrangements in the brain, even in an adult one, incline me towards a non-rigid and functional hypothesis of the "regulation of access" to the different representation systems. This is in agreement with the tendency place increasing importance on the cerebral plasticity, especially in very early infancy. This is also demonstrated in cases of extensive malformations of one hemisphere requiring early radical hemispherectomy where we can see how the right hemisphere almost completely takes over all the functions of the other hemisphere, privileging those higher functions that are essential from the adaptive point of view (Canestri 2000). Without insisting too much on this issue, from this example we are able to draw some conclusions about the innatist hypotheses that privilege excessively rigid regional localizations or specializations.

The same amount of attention should be paid to the aspects that are connected to the development of the entire linguistic system. While the phonological and syntactical aspects linked to the unaware procedural memory are already well developed before the age of three, in harmony with the advanced maturation of the competent nervous structures (basal ganglia, cerebellum, sensory-motor cortical areas), the lexical aspects of language come later, in correspondence with the later maturation of the nervous structures linked to the declarative memory (hippocampus and temporal-parietal areas). In the same way the auditive development and corresponding perception is much more rapid than the development of the phonatory articulation system (three months for the first against eighteen months for the second). Auditory reactions and recognition occur very early (even in the uterus and mainly concerning the mother's voice). Unlike the phonatory articulation system, the auditory system needs no practice and is a system that is specialized right from the start for accomplishing a sensorial function, while the phonatory system is the heir to a long adaptive history that has transformed a visceral system (pneumogastric nerves) into a system with a linguistic purpose (Hagège 1996). I mention these aspects in order to emphasize the importance that a careful consideration of the concurring elements, with different maturating times and modalities, can have in the development of the "linguistic system" in the child. From the psychoanalytical angle as well, it is impossible to overlook the fact that the moments of psychological development, which are so crucial in our discipline, play a significant role in determining the quality of the representations and of the words to which they are linked. I will briefly return to the role of pragmatics in language acquisition that I mentioned earlier.

A pertinent aspect in the study of linguistic development in the mono or bilingual child is certainly that of the "linguistic" exchange with the environment and, obviously in a privileged manner, with the mother. Putting "linguistic" in inverted comas indicates that this concept must be understood, in this case, in its widest sense and close to that of a "complex semiotic system", some of whose elements are strictly linguistic and others not. It is at this level that the learning of the "mother tongue" meets the primary relationship with the mother, even though the two must not be homologated. It would be misleading today to study language acquisition without mentioning the so-called "baby talk, motherese or child directed speech", i.e. without taking into consideration the ways in which we speak *with* babies.

I have intentionally emphasized *with* babies because, as Snow (1986) pertinently writes, mothers do not speak *to* babies but *with* them, in a manner that could be qualified as *conversational* (and in the very early phases of life as *protoconversational*) and that is governed by the child. The mother does not try to teach her child to speak, but tries to effectively communicate with him, and to do this she uses certain particular modalities that have been amply described, analyzed and discussed. What is important to consider is the fact that the learning processes perhaps differ in the different periods of the development of linguistic acquisition, that there are more or less efficacious processes of linguistic adaptation between adults and children, that there exist some pathological (non-answer of the mother to the child's expressions, answers that are unconnected and not semantically or affectively pertinent: intonation, modulation, rhythm, etc.) "dyadical constructions" (E. Lieven 1978) that in psychoanalysis we would consider as a consequence of the pathology of the mother-child relationship, and that the entire linguistic learning process is intimately linked to the affective "monitoring" between mother and child and more widely between child and environment. The "pathological dyadical constructions" do not seriously hinder the learning of language as such (at the most they could slow it down, favoring a language that is poor, repetitive, not very informative or difficult to interpret, consequently impoverishing the feedback and increasing the problem of communication), but it is evident that they can be at the root of many difficulties described in clinical cases and are psychoanalytically relevant. An increase of studies in this direction would be advisable. The simultaneous presence of several languages and the relationship that is installed with them in the child's primary environment could alternatively increase his creativity and affective-linguistic richness, or induce an additional pathology if the communication system is already pathological. This last eventuality, therefore, would not depend on the presence of several languages, as we have abundantly tried to demonstrate (Amati Mehler, Argentieri & Canestri 1993), but on a defect of the system.

At the beginning of this paragraph, when speaking about linguistic exchange with the environment, I said that the word "linguistic" is to be understood in the sense of a complex semiotic system. This is justified not only because during linguistic interaction the child, until he is eighteen months old, is in a pre-linguistic phase (or pre-verbal, if you prefer), but also because linguistic communication is inseparable from gestural and motorial expressions, from prosodic levels and from the use of multiple channels of communication that are a part of the system: physical contact, glances, etc. Regarding this, it is enlightening to observe how the spontaneous construction of sign "languages" in deaf children follows the language acquisition programme.

The theme of affective mother/environment-child exchange has strong connections with what I have said above. Whatever the preferred psychoanalytical theory may be, we can all agree on the words of R. Pally (1998): "Emotion connects not only the mind and the body of one individual but minds and bodies *between* individuals" (p. 349). Whether or not one agrees with his meta-theories, C. Trevarthen (1977, 1979) in his pioneeristic works emphasizes, with precise reference to the pre-linguistic mother/environment-child exchange, the decisive role played by affect attunement and the significance of the reciprocal affective monitoring in providing the means for human communication and the acquisition of language and of symbolic capacities.

One example only could be sufficient to illustrate it. The children of deaf-mute parents who needed linguistic stimuli in order to learn to speak – inasmuch as language acquisition is a good example of the nature-nurture interaction – were entrusted for a period of time to the linguistic "care" of a video recording. None of them learnt to speak, not only because, as S. Pinker (1995) states, the video recording did not take into consideration the here-and-now of the discourse, or allow to analyze the wish to say more or other than what is said (a pragmatic aspect that is fundamental in verbal communication), but above all because no affect attunement or affect monitoring was possible with the machine.

The argument I am considering, as I anticipated when listing the different issues to be taken into account, also includes the theme of the relationships between thought and language. It would require another paper, and I have included it in the list only as a reminder that in this field also, psychoanalytic theory would benefit from a conceptual update. In a book written in 1998, *Language and Thought*, as the result of an interdisciplinary project, the editors of the volume – Peter Carruthers and Jill Boucher – wrote "we felt that the issue of language and thought had almost dropped from sight in the cognitive sciences in recent decades, and that the time was ripe for a more fine-grained examination of the issues from an interdisciplinary perspective" (p. XIII). The same observation is valid for psychoanalysis; a reflection on the problems relative to the inner language and to the different links between thought and language, between types of representations, etc., would be useful.

5. GENERAL THEORY OF REPRESENTATION

The French linguist A. Culioli, who I have already mentioned, has outlined three different levels in the treatment of representation. The first is the level of mental representations, whose conceptualization depends on the discipline involved. The access to this level is obviously not direct. The second level is that of linguistic representations: they carry traces of the activity of the first level, but we cannot hypothesize a bi-univocal relation between the two, due to the fact that representations are heterogeneous. The third level is that of meta-linguistic representations. It is reasonable to presuppose a level of neural activity which precedes and informs the first level and that could be conceived also as heterogeneous if confronted with the others. Freudian psychoanalytical theory introduced a distinction between thing presentation (Dingvorstellung) and word presentation (Wortvorstellung) that cannot be assimilated to the described oppositions. Greatly discussed and criticized from various points of view (e.g. the above mentioned work by Litowitz), but also greatly defended as one of the fundamental concepts of psychoanalysis (e.g. A. Green 1997, 2000, and his theories on tertiary processes), the Freudian concept cannot easily be substituted without impairing some of the nuclear aspects of the theory itself. It could, however, be reformulated: one of these reformulations was made by Green himself and is based on Peirce's theory of the sign (representamen). Following in the wake of his hypotheses, a general theory on representation in psychoanalysis can perhaps be outlined.

It could and should take into account the classification of the interpretants into Immediate, Dynamical and Final that Peirce traced in 1908 within the framework of his general theory on semiosis. Let us recall Peirce's well-known definition on what a sign (Representamen) is, and what, therefore, semiosis is. Peirce says "A Representamen [sign] is the subject of a triadic relation with a second, called its Object, and a third, called its Interpretant. This triadic relationship being such that the Representamen determines its interpretant to stand in the same triadic relation to the same object for any Interpretant" (Peirce 1903, p. 285 [1.541]). "Peirce's theorization on semiosis evolves in the form of a dynamic and substitution process. It is a process that includes a very wide range of intra-and-extra-linguistic material to be included in semiosis (feelings, actions, body signals, every class of sign...)" (Canestri 2002, p. 536). However the process *must include* the final interpretant. The final interpretant is the manner in which the sign tends to represent itself in relation to its object, or the effect that the sign produces in the mind after a sufficient development of thought, even in the mind of the interlocutor when, for example, you ask a question or enounce any kind of utterance. A good illustration is certainly given by Freud's joke (Peirce 1908, p. 232 [8.343]).

In this paper many other perspectives, as well as many other works, authors and lines of research, would have "right of entry". I have chosen only a few samples of what the sciences of language have to offer to psychoanalysis. Of course, I could attempt to carry out the operation in reverse, and I think this would be just as fruitful: I could suggest to the sciences of language what psychoanalysis has to say to them.

REFERENCES

Amati Mehler, J., Argentieri, S. & Canestri, J. (1993). *The Babel of the Unconscious*. Madison, CT: International Universities Press.

Apel, K.O. (1986). *Le logos propre au langage humain*. Combas: Ed. de l'Eclat 1994.

Austin, J.L. (1962) *How to Do Things with Words*. Oxford: Clarendon Press.

Bakhtin, M. (1978) *Esthétique et théorie du roman*. Paris: Gallimard.

Canestri, J. (1991). La passione della lingua originaria nella letteratura e nella psicoanalisi. *Il Piccolo Hans* 69: 160–186

Canestri, J. (1993). Il soggetto polilogico. In *Ri-scritture,* ed. Centro Romano di Semiotica. Rome: Eurelle Edizioni.

Canestri, J. (1994a). From internal flux to shared experience. In *The Analyst Mind: From Listening to Interpretation*. London: The IPA Press, pp. 23–30.

Canestri, J. (1994b). Transformations. *Journal of the American Psychoanalytic Association* 75: 1079–1092.

Canestri, J. (1999a). Alcune note e un esempio su attività linguistica e psicoanalisi. *Psicoanalisi* 3(1): 68–82.

Canestri, J. (1999b). Development of affect in bilingual patients. Presented at the 41st International Psychoanalytical Congress, Santiago de Chile.

Canestri, J. (2000). Notes on linguistic activity and psychoanalysis. In *Changing Ideas in a Changing World. The Revolution in Psychoanalysis*. London and New York: Karnac Books, pp. 197–205.

Canestri, J. (2001). Ferenczi e l'attività linguistica nell'analisi. In *La catastrofe e i suoi simboli*. Torino: UTET, pp. 187–197.

Canestri, J. (2002). Trois, le nombre magique: Green avec Peirce. In *Penser les limites. Écrits en l'honneur d'André Green*. Paris: Delachaux et Niestlé, pp. 531–538.

Carruthers, P. & Boucher, J., ed. (1998). *Language and Thought. Interdisciplinary Themes*. Cambridge: Cambridge University Press.

Charaudeau, P. & Maingueneau, D. (2002). *Dictionnaire d'analyse du discours*. Paris: Seuil.

Chomsky, N. (1986). *Knowledge of Language. Its Nature, Origin and Use*. New York: Praeger.

Chomsky, N. (1995). *The Minimalist Program*. Cambridge, Mass.: MIT Press.

Chomsky, N. (2000). *New Horizons in the Study of Language and Mind*. Cambridge: Cambridge University Press.

Culioli, A. (1990). *Pour une linguistique de l'énontiation. Operations et représentations. Tome I*. Paris: Ophrys.

Culioli, A. (2002). *Variations sur la linguistique*. Paris: Klincksieck.

Danon-Boileau, L. (1987). *Le sujet de l'énonciation. Psychanalyse et linguistique*. Paris: Ophrys.

Davidson, D. (1984). *Inquiries into Truth and Interpretation*. Oxford: Oxford University Press.

Diebold, A.R. Jr. (1989). The consequences of early bilingualism in cognitive development and personality formation. In *On the Bilingual Person*, ed. R. Titone. Ottawa, Ontario: Canadian Society for Italian Studies, pp. 65–92.

Elman, J.L., Bates, E.A., Johnson, M.H., Karmiloff-Smith, A., Parisi, D. & Plunkett, K. (1996). *Rethinking Innateness: A Connectionist Perspective on Development*. Cambridge, Mass.: MIT Press.

Fabbro, F. (1999). *The Neurolinguistics of Bilingualism*. Howe: Psychology Press.

Fónagy, I. (1983). *La vive voix. Essais de psycho-phonétique*. Paris: Payot.

Freud, S. (1905c). *Jokes and their Relation to the Unconscious*. S.E. 8.

Gopnik, A. & Meltzoff, A.N. (1997). *Words, Thoughts and Theories*. Cambridge: Cambridge University Press.

Gopnik, A. (2000). Theories, language and culture: Whorf without wincing. In *Language Acquisition and Conceptual Development*, ed. M. Bowerman & S. Levinson. Cambridge: Cambridge University Press.

Green, A. (1984). Le langage dans la psychanalyse. In *Langages: Rencontres psychanalytiques d'Aix-en-Provence*. Paris: Les belles lettres, pp. 20–250.

Green, A. (1997). Le langage au sein d'une théorie générale de la représentation. In *Pulsions, représentations, langage*, ed. M. Pinol-Douriez. Laussanne: Delachaux et Niestlé, pp. 23–66.

Grice, H.P. (1989). *Studies in the Way of Words*. Cambridge, Mass.: Harvard University Press.

Hagège, C. (1985). *L'homme de paroles*. Paris: Fayard.

Hagège, C. (1996). *L'enfant aux deux langues*. Paris: Editions Odile Jacob.

Kail, M. & Fayol, M. (2000). *L'acquisition du langage.* Paris: Presses Universitaires de France.

Karmiloff-Smith, A. (1992). *Beyond Modularity: A Developmental Perspective on Cognitive Science.* Cambridge, Mass.: MIT Press/Bradford Books.

Kim, K.H.S., Relkin, N.R., Lee, K.M. & Hirsch, J. (1997). Distinct cortical areas associated with native and second languages. *Nature* 388: 171–174.

Lacan, J. (1966). *Écrits.* Paris: Seuil.

Liberman, D. (1983). *Linguística, interacción comunicativa y proceso psicoanalítico.* Buenos Aires: Ediciones Kargieman.

Lieven, E. (1978). Conversations between mothers and young children: individual differences and their possible implications for the study of language learning. In *The Development of Communication*, ed. N. Waterson & C.E. Snow. Chichester: Wiley, pp. 173–187.

Lieven, E. (1978). Turn-taking and pragmatics: two issues in early child language. In *Recent Advances in the Psychology of Language: Language Development and Mother-Child Interaction*, ed. R.N. Campbell & P. Smith. New York & London: Plenum Press.

Lieven, E. (1979). Communication and cooperation in early infancy. A description of primary intersubjectivity. In *Before Speech, The Beginning of Human Communication*, ed. M. Bullowa. London, pp.321–347.

Litowitz, B.E. & Litowitz, N. (1977). The influence of linguistic theory on psychoanalysis. A critical, historical survey. *International Review of Psychoanalysis* 4: 419–448.

Mergenthaler, E. & Bucci, W. (1999). Linking verbal and non-verbal representations: Computer analysis of referential activity. *British Journal of Medical Psychology* 72: 339–354.

Pally, R. (1998). Emotional processing: the mind-body connection. In *The Mind-Brain Relationship.* London & New York: Karnac Books, 73–104.

Paradis, M., ed. (1995). *Aspects of Bilingual Aphasia.* London: Pergamon Press.

Peña, M., Bonatti, L.L., Nespor, M. & Mehler, J. (2002). Signal-driven computations in speech processing. *Science* 298: 604–607.

Peirce, C.S. (1903). Degenerate cases. In *Collected Papers of Charles Sanders Peirce*, ed. C. Hartshorne & P. Weiss. Vol. 1. Cambridge, Mass.: Harvard University Press 1931, pp. 277–286.

Peirce, C.S. (1908). To Lady Welby. In *Collected Papers of Charles Sanders Peirce*, ed. C. Hartshorne & P. Weiss. Vol. 8. Cambridge, Mass.: Harvard University Press 1931, pp. 220–245.

Pinker, S. (1995). *The Language Instinct.* New York: Harper Perennial.

Searle, J.R. (1969). *Speech Acts.* London: Cambridge University Press.

Shapiro, T. (1988). Language structure and psychoanalysis. *Journal of the American Psychoanalytic Association* 36: 339–358.

Shapiro, T. (1991). Words and feelings in the psychoanalytic dialogue. *Journal of the American Psychoanalytic Association* 39: 321–348.

Snow, C.E. (1986). Speaking with children. In *Language Acquisition*, ed. P. Fletcher & M. Garman. Cambridge: Cambridge University Press, pp. 105–132.

Sperber, D. & Wilson, D. (1986). *Relevance.* Oxford: Basil Blackwell.

Trevarthen, C. (1977). Descriptive analyses of infant communication behaviour. In *Studies in Mother Infant Interaction*, ed. H.R. Schaffer. London: The Loch Symposium, pp. 227–270.

Vinocur de Fischbein, S. (1998). Psicoanálisis y lingüística: contactos e intercambios. Unpublished manuscript.

AUTHORS

Brakel, Linda, A.W., Dr. (USA)
Clinical Assistant Professor of Psychiatry, Department of Psychiatry, Assistant Director of the Ormond and Hazel Hunt Event-Related Potential Laboratory, University of Michigan Medical School; Psychoanalyst: Member and Faculty Member of the Michigan Psychoanalytic Institute, American Psychoanalytic Association (APsA)/International Psychoanalytical Association (IPA).

Canestri, Jorge, Dr. (I)
Professor of Psychiatry, Department of Psychiatry, University of Rome; Psychoanalyst: Member, Training and Supervising Analyst of the Italian Psychoanalytical Association (AIPsi) and of the Argentine Psychoanalytic Association (APA)/International Psychoanalytical Association (IPA).

Giampieri-Deutsch, Patrizia, Dr. (A/I)
Universitätsdozent at the University of Vienna; Psychoanalyst: Member, Training and Supervising Analyst of the Viennese Psychoanalytical Society (WPV)/International Psychoanalytical Association (IPA), Research Fellow of the International Psychoanalytical Association (IPA), Chair of the Ethics Committee of the Viennese Psychoanalytical Society (WPV/IPA).

Hauser, Stuart, Dr.Dr. (USA)
Professor of Psychiatry at the Harvard University Medical School, President of the Judge Baker Children's Center of the Harvard University Medical School; Psychoanalyst: Member, Training and Supervising Analyst of the Boston Psychoanalytic Institute of the American Psychoanalytic Association (APsA)/International Psychoanalytical Association (IPA).

Jacob, Pierre, Dr. (F)
Professor of Philosophy at the Ecole Polytechnique (CREA) and at the Institut des Sciences Cognitives de Lyon, Director of the Institute Jean Nicod at the Centre National de la Recherche Scientifique (CNRS), Member of the Committee of the European Society for Philosophy and Psychology.

Levin, Fred, M., Dr. (USA)
Assistant Professor of Clinical Psychiatry, Northwestern University School of Medicine, Member of the Neurological Faculty, University of Chicago Medical School; Psychoanalyst: Member, Training and Supervising Analyst of the Chicago Institute for Psychoanalysis and the Minneapolis Psychoanalytic Institute, American Psychoanalytic Association (APsA)/International Psychoanalytical Association (IPA).

Pally, Regina, Dr. (USA)
Professor of Psychiatry at the University of California (Los Angeles); Psychoanalyst: Member, Training and Supervising Analyst of the Los Angeles Psychoanalytic Society, American Psychoanalytic Association (APsA)/International Psychoanalytical Association (IPA).

Pribram, Karl, Dr. Dr. h. c. mult. (USA)
Professor of Psychology and Cognitive Science, Director of the Faculty of Cognitive Science, Georgetown University Washington, Professor Emeritus of Neuroscience in the Departments of Psychology and Psychiatry, University of Stanford, James P. and Anna King Distinguished Professor, Radford University and Adjunct Professor at the University of Tennessee at Knoxville.

Rosenthal, David, Dr. (USA)
Professor of Philosophy at the City University of New York and Clark-Way Harrison Visiting Professor at the Washington University, St. Louis, Director of the Program of Cognitive Sciences at the City University of New York.

Sandell, Rolf, Dr. (S)
Professor of Psychology, Department of Psychology and Pedagogy, Linköping University; Psychoanalyst: Member, Supervising and Training Analyst of the Swedish Psychoanalytical Society (SPaF)/International Psychoanalytical Association (IPA), Research Fellow of the International Psychoanalytical Association (IPA).

Shevrin, Howard, Dr. (USA)
Professor of Psychology, Department of Psychology and Psychiatry, University of Michigan Medical School, Director of the Ormond and Hazel Hunt Event-Related Potential Laboratory; Psychoanalyst: Member and Faculty Member of the Michigan Psychoanalytic Institute,

American Psychoanalytic Association (APsA)/International Psychoanalytic Association (IPA).

Velmans, Max, Dr. (GB)
Professor of Psychology, Department of Psychology, Goldsmith College of the University of London.